DIGITAL IC PROJECTS

by
F.G. RAYER, T.Eng.(CEI), Assoc. IERE

BERNARD BABANI (publishing) LTD
THE GRAMPIANS
SHEPHERDS BUSH ROAD
LONDON W6 7NF
ENGLAND

Although every care has been taken with the preparation of this book, the publishers or author will not be held responsible in any way for any errors that might occur.

©1981 BERNARD BABANI (publishing) LTD

First Published — March 1981

British Library Cataloguing in Publication Data
Rayer, Francis George
 Digital IC projects.
 1. Integrated circuits — Amateurs' manuals
 I. Title
 621. 381'. 73   TK 9965

ISBN 0 85934 059 7

Printed and bound in Great Britain by Cox & Wyman Ltd, Reading

CONTENTS

	Page
INTRODUCTION	1
COMPONENTS	2
ICs and Holders	2
Alternative Numerals	3
GN4 Nixie	4
7-Sector LEDs	5
Other Components	7
POWER SUPPLIES	9
Battery Running	9
Transistor Regulation	11
AC Supply	12
IC Regulator	14
Nixie Supply	15
Multiplier for Nixie Supply	16
PROJECTS	18
Nixie Numerator	18
1-6 Numerator	20
"Steady Hand" with Counter	24
"Twinkle Tree"	27
Roulette	29
Six-Spot	32
Noiseless Switch	38
Testing BCD and Decoder-Driver	39
LED Random Numerator	42
Scorer for Bezique, Etc.	46
Multi-Digit Counter	54
10X Dividers	55
Digital Stop-Clock	56
Quartz Stop-Clock	60
Response Time Indicator	67
Sound Initiated Timer	68
Light Operated Counters	70
6-Digit Frequency Counter	73
Digital Signal Generator	81

	Page
Radio Frequency Marker	84
Variable 555 Pulser	88
1-Armed Bandit	89

INTRODUCTION

This book contains both simple and more advanced projects, and it is hoped that these will be found of great help in developing a knowledge of the workings of digital circuits.

Various forms of assembling and wiring the integrated circuits on their boards are shown, and this aspect of a project can be quite straightforward, and not need the preparation of a printed circuit board. The more ambitious projects can be built and tested step by step, and this will avoid or correct faults which could otherwise be troublesome, and will result in a better understanding of how the devices operate.

COMPONENTS

ICs and HOLDERS

The integrated circuits fit boards of 0.1in matrix, and ICs and their holders have the same pin spacings, though the holders are a little wider than the ICs they carry. Actual holder designs vary somewhat.

Pin 1 may be indicated by a notch, depression, dot, number 1, or ridge, Figure 1. The holders generally have some means of showing socket 1, so that ICs may be inserted the correct way round. When an IC is to be inserted for the first time, all its pins may need pressing inwards very slightly. Any pins obviously bent in packing should be carefully straightened with tweezers or similar means.

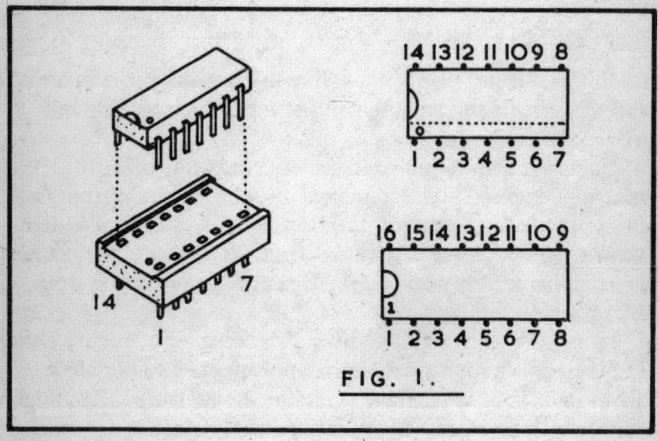

FIG. 1.

Most ICs required here have 14 or 16 pins, counted as in Figure 1 when viewing the IC from the top. ICs can be soldered directly in place, but the use of holders allows their easy replacement, or substitution for test purposes.

The 74 TTL (transistor-transistor logic) ICs used require 4.75V to 5.25V and normally operate from a 5V supply. Unused gate inputs should be taken to negative, or may be connected to a used input, or may be connected to positive through a resistor.

The NAND function is as follows:

Inputs	Output
0.0	1
0.1	1
1.0	1
1.1	0

The binary coded decimal decade counter counts from 0 to 9 giving binary outputs 0000 to 1001 at four pins. A decoder-driver decodes these outputs into the form required to display numbers 0 to 9 on an associated numeral. Details appear in the related sections.

An inverter inverts the logic, *low* input giving *high* output, and *high* input providing *low* output.

ALTERNATIVE NUMERALS

Counting and similar devices will require one or more numerals, and the circuits shown here will allow 7-segment LEDs and Nixie tubes to be employed.

Numerals with illuminated sectors consisting of light emitting diodes, or LED numerals, may be operated from the 5V supply used to power integrated circuits. Nixies, however, require a high voltage supply, though this can be of very simple type, and is readily obtained by transformer operation from AC mains.

In many of the circuits shown, it is feasible to employ either LED or Nixie numerals, with the appropriate LED or Nixie driver ICs. Thus as example, a circuit shown with LEDs, might equally well be made with Nixies, in most cases.

Numitron and Minitron numerals, having small filamentary sectors, are also often seen, but have not been dealt with here. End and side viewing Nixie numerals, Numitrons and Minitrons are from time to time seen on sale singly or in packs at low cost, and this can be a factor in adopting them for a display.

Nixie circuits in this book are for the GN4 (and equivalent) tubes. The GN4 is end-viewing, with numerals about 15mm high. Side viewing Nixies may be substituted, and would need to stand vertically at the edge of the circuit board, or be fixed with the numeral side flat to the viewing aperture.

Other details appear in book number BP67 Counter Driver and Numeral Display Projects also published by Bernard Babani (publishing) Ltd.

GN4 Nixie

This has 13 pins, shown in Figure 2. Counting from the space, pin 2 is the connection to the current limiting resistor R1. This can be 33k or 47k ½ watt for a 240V to 250V supply. Current drawn is around 1½mA to 3mA. Higher values of series resistor, or lower HT supply voltage, will reduce brightness, but these factors are not critical.

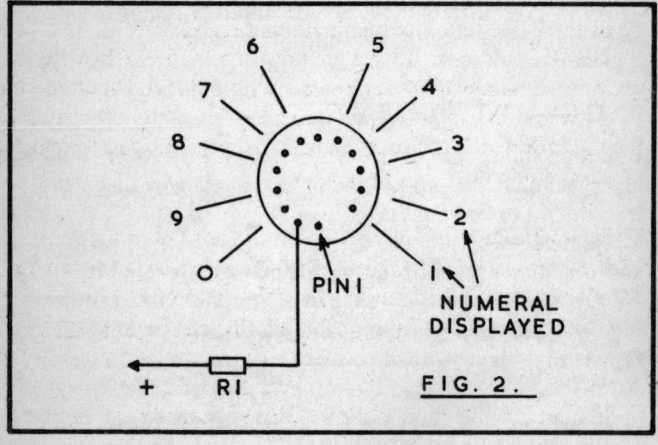

The required numeral is illuminated by taking another pin to negative. Pin 3 used in this way produces 0, pin 4 provides 9, pin 5 illuminates the numeral 8, and so on. Pins 1 and 8 provide decimal type points in some tubes, and should have a further series limiting resistor (about 220k). Surplus or ex-equipment numerals of this kind are readily tested by clipping R1 to pin 2, and stepping round the other pins with a lead returned to negative, observing numerals produced. Very aged, much-used tubes may produce part numerals, but might if to hand be suitable for positions where a limited display (such as only 0–1, or 0–1–2 for clocks) will be needed. No display at all probably means lost seal due to distorted pins.

These tubes may be inserted in holders, or connections can be taken directly to the pins, and in some ways this has advantages. All pins should be clean and bright. Use a thin flexible or single-strand bright tinned-copper wire, with sleeving a fairly tight push-fit on the pins. Bare a short length of the wire, insert in the sleeving, and push on the pin. The leads can be as long as needed to reach from the board carrying the ICs.

Nixies are operated from driver ICs which decode the binary input into a single circuit output. Thus when the driver IC for the Nixie receives the binary equivalent of 1, it automatically gives a negative connection to a pin which is connected to pin 13 of the Nixie, thus illuminating 1. When the binary changes to 2, the IC alters the circuit, so that this is completed to pin 12 of the Nixie tube, to illuminate numeral 2, and so on.

The Nixie decoder-drivers are not used for LED displays, as with these, two or more segments are illuminated simultaneously.

Connections can be arranged to provide counts other than the 0–9 for which the tube is made, to suit the need for clocks or other equipment.

7 – Sector LEDs

Figure 3 shows a LED numeral with 7 segments. Two or more are illuminated, to produce numerals as follows:

0	ABCDEF	5	AFGCD
1	BC	6	AFEDCG
2	ABGED	7	ABC
3	ABCDG	8	ALL SECTORS
4	BCFG	9	GFABCD

With many circuits 6 omits A and 9 omits D.

The individual sectors are separate light emitting diodes, and all these LEDs are connected together at one end. Where the decoder-driver is placed to control the negative circuits, as in circuits here, common anode LEDs are required. This common positive connection is pin 3; or pins 3, 9 and 14 with some types.

The number of segments illuminated, and drawing current, will depend on the number shown, there is a minimum of 2

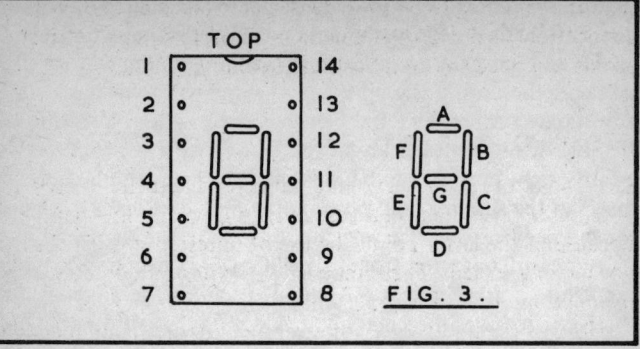

FIG. 3.

sectors for number 1, and maximum of 7 sectors for number 8. Current taken by the numeral thus varies considerably, and a common limiting resistor is not very satisfactory. For this reason, each sector can best have its own resistor, between the driver IC and the sector pin.

Referring to Figure 3 pins and segments are:

    A pin 1   E pin 7
    B pin 13   F pin 2
    C pin 10   G pin 11
    D pin 8   Common, pin 3.

Thus, to illuminate BC for 1, circuits are completed to pins 13 and 10. To obtain 0, circuits would be completed to all sectors listed except G, pin 11.

The decoder-driver ICs used receive the binary input in the usual form, for 1, 2, 3 and so on, and automatically provide output circuits for the correct segments. These drivers cannot of course be used with Nixies, for the reason explained.

LED numerals are available in various sizes and colours, and to plug into 14 pin dual-in line holders. The DL707 has 0.3in high numbers, fits 14 pin DIL holders, and is suitable for circuits here.

Testing

From the foregoing, it will prove quite straightforward to test a numeral which is not giving a correct display. If the numeral indicates correctly when appropriate circuits are made to it, then the fault has to be sought elsewhere (such as in driver connections). But if the numeral is found to be faulty, this avoids any need for investigation of other circuits.

OTHER COMPONENTS

The transistors and integrated circuits used here are easily obtainable from many sources, and are generally inexpensive. Because of the latter point, the use of untested surplus ICs is not recommended. These may have one or more defects, troublesome to locate.

Holders for the ICs are not essential, but are recommended for all positions. They are Dual In Line (DIL), mostly 14 pin and 16 pin, and are available in various styles.

If an IC is to be easily inserted in a holder for the first time, it is usually necessary to bend its pins slightly to make them parallel. This can be done with small pressure. Without care at first insertion, a pin may bend, or not engage correctly with its socket. Check each IC is the correct way round, as it is put in.

An IC can be removed from its socket by inserting a thin blade between socket and IC, and levering it up a little at each end. Continue until it is free.

In some circuits, operation will cease with a quite small drop in voltage, so that a correct supply must be available. Other circuits are less liable to fail in this way. Some will operate partially, giving results which could suggest obscure faults in counting or other circuits.

Excess voltage, or wrong polarity, must be avoided. Full details of various means of supplying power are provided, and a small mains operated power supply unit will be of great utility.

Resistors can generally be of ¼ watt and smaller sizes, and these are easiest to accommodate. Capacitors, except for a few isolated positions, are low voltage. Electrolytics may be 6V, 6.4V, 9V, 10V, 12V or as conveniently available. These are generally for by-pass purposes, so that values are not very critical. Similarly, in the smaller values of non-electrolytic capacitor, 47nF and 0.05 μF (50nF) are interchangeable. But in timing or oscillator circuits, values should be as given.

Boards are 0.1in matrix, as the ICs or holders fit these. The various methods of wiring shown generally require boards without foil. If holders, etc., are placed one side of the board, and wiring is largely the other, rapid assembly is possible, with this method. Some units show wires run above and below the board, to obtain cross-overs. This will generally require a larger board.

For soldering on 0.1in matrix boards, a small iron, typically 15 watt with bit to suit, will be almost essential. This, with a good quality cored solder, will make soldering easy. A good light is helpful when working, and occasionally a hand magnifier to examine some connections may prove useful.

Some layouts show boards with underneath connections vertical or horizontal. Board foils can be used for these circuits, foils elsewhere being cut with a drill or spot face cutter. Foil clad board may also be used with point-to-point wiring, if wished, by cutting foils each side of connecting points. Foil clad board can also be cleared by immersing in etching solution as if making a printed circuit; or by means of a file, for small boards, giving support by a flat surface, or by use of a power abrasive disc, preferably out-of-doors.

Low voltage switches can be of the small slide type; or can be miniature or standard size toggles.

Where a single power supply unit is going to be employed from time to time with various units, each may have a flex with non-reversible plug, to match a socket on the PSU. Small, low-voltage plugs and sockets are available for this purpose.

Stout leads are unnecessary, and introduce troubles such as broken pins, or short-circuits at joints. Most connections can be of thin wire, such as 30swg tinned copper. A stouter wire is used for positive and negative lines, where several ICs are present.

It is very useful to have a few different colours of narrow gauge sleeving, to put on leads, and help identify these. Very thin flex is also best for some other purposes, red for positive, black for negative, and other colours for switching circuits, etc. However, a non-miniature flexible conductor should be used for the main power supply circuits, unless these are very short, and only two or three ICs are present.

POWER SUPPLIES

Current will most generally be derived from AC mains, a reliable and economical supply being obtained by this means. However, for some circumstances, battery running is feasible.

Operation from dry batteries will generally be for items which require only a relatively small current, or are of a type operated for short intervals. Battery running can also be adopted in some other circumstances, as when no mains will be available, or when mains may not be used, or when an adequate current supply is available, as in a vehicle.

It is impossible to state how many hours' use will be obtained from any particular battery pack, but some idea may be gained by inserting a meter in one supply lead to note average current drawn. This can be compared with that taken by the typical hand-lamp, probably fitted with a 0.3A bulb.

BATTERY RUNNING

The 7400 series of ICs normally operate from about 5V, with lower and upper limits of 4.75V and 5.25V, and an absolute maximum of 5.5V. It will be found that many devices can be run from a 3-cell or 4.5V battery. This has the advantage of needing no series dropper or other voltage control.

Where the full 5V will be necessary, and current drain is reasonably steady, it is in order to use a 4-cell battery pack, for 6V, and drop the excess 1V with a series resistor.

Where the load varies throughout operation, as it often does, the straightforward series limiting resistor is not suitable. It is then best to employ a Zener diode for regulation, or transistor or IC regulators. With these, there can be an advantage in having a rather higher voltage available, such as 7.5V, 9V or 12V.

The situation for a steady current is shown at A, Figure 4. A 6V supply is reduced to 5V by R1, which drops 1V. From Ohm's Law, $R = V/I$ (voltage divided by current). So if 100mA will be required, 1 volt must be dropped at 0.1 ampere, 1/0.1, so R1 is 10 ohm. Changes in current drawn will result in corresponding changes in the voltage lost in R1, so that the supply does not remain at 5V.

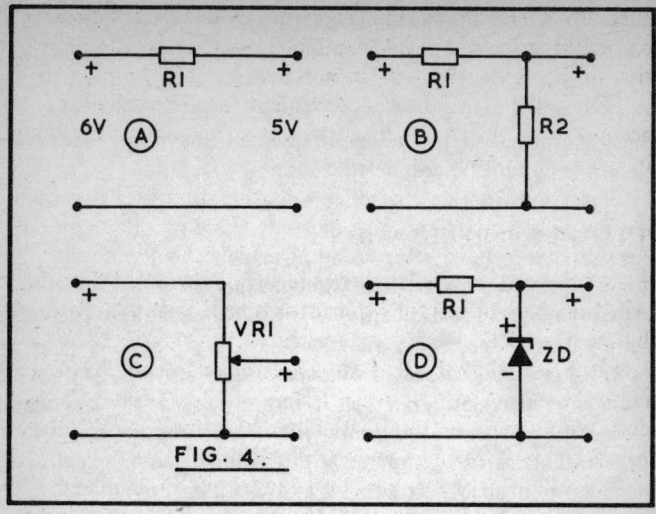

FIG. 4.

The circuit at B is slightly better, as constant current drawn by R2 helps swamp changes in current drawn by the equipment so the voltage dropped in R1 is subject to smaller variation. The larger the current drawn by R2, the greater the improvement. This is practical where a simple mains unit providing 1A or more for R2 can be brought into service, and a 10 ohm wire wound potentiometer may be connected as in C, and adjusted for 5V output at minimum load. It is also a temporary expedient for accumulator running.

D uses a Zener diode. Assuming a 12V supply, R1 must drop 7V, to leave 5V available. With a 5V ZD rated at 1 watt, the ZD current cannot exceed 200mA, thus R1 is 35 ohm 1½W. In this case, an output of 0–200mA can be drawn. Should the minimum load to be provided be known, this can be allowed for, to up-rate the maximum load. As example, assume the minimum load at 5V is 200mA (not zero). With 200mA flowing through the diode, the total through R1 is 400mA, so it may be reduced to 18 ohm 3W. Maximum current which may be drawn will then be about 400mA.

Where R1 is too low in value for the current the ZD must pass at minimum external loading of the supply, the diode current is too great, and it is damaged. Where external current exceeds that for which R1 is suitable, the voltage falls below that of the diode, and regulation is lost.

Zener diode regulation is convenient for some vehicle or accumulator operated equipment, where the continuous drain of the diode is of no importance.

TRANSISTOR REGULATION

Figure 5 shows a transistor series regulator for 12V DC input. At A, the base of the NPN transistor is kept at a stable voltage by the diode ZD1, which receives current through R1, 270 Ω ¼ watt. A 400mW diode is adequate here, as it only has to control the base current of the transistor. The emitter voltage, and consequently output, is stabilised at near the base voltage, for relatively large changes in current drawn.

A small difference in potential exists between base and emitter, and may be around 0.6V. ZD1 can thus be 5.6V, for a 5V output. An output test with a meter should be made before connecting equipment.

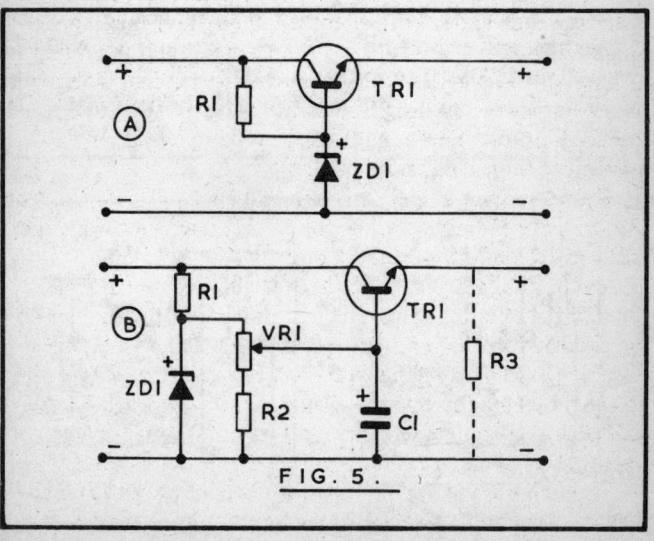

FIG. 5.

Circuit B allows adjustment of output voltage. ZD1 provides a slightly higher voltage than required, and VR1 allows the base of the control transistor (and output voltage) to be set, with the aid of an accurate, high-resistance voltmeter. ZD1 may be 6V or of higher voltage. VR1 can be around 500 ohms, or may be of lower value, with R2 in series, to open out adjustment. R1 is reduced to 150 ohm, ½ watt. R3 may be provided to draw a small current when no external load is present. R3 can be 470 ohm.

C1 can be added for an AC operated supply, for electronic smoothing of output via the transistor, and can be 100 μF, 10V

Various NPN audio and output type transistors may be used, to suit the maximum current likely to be required. The BD139 is suitable for 1.5A maximum, and for larger current the 2N3055 is unlikely to be exceeded. A chassis, panel, case or heat-sink should be utilised as mounting, to keep temperature down, except for low currents where this is unnecessary.

AC SUPPLY

Figure 6 shows means of obtaining power from AC mains. Circuit A has a centre-tapped transformer. With this method, C1 will charge up to just over 1.4 times the indicated rectifier connecting points. So with a 9—0—9V secondary, C1 will reach about 12.6V. The voltage across C1 will fall, as current is drawn. When this has dropped to the level where ZD1

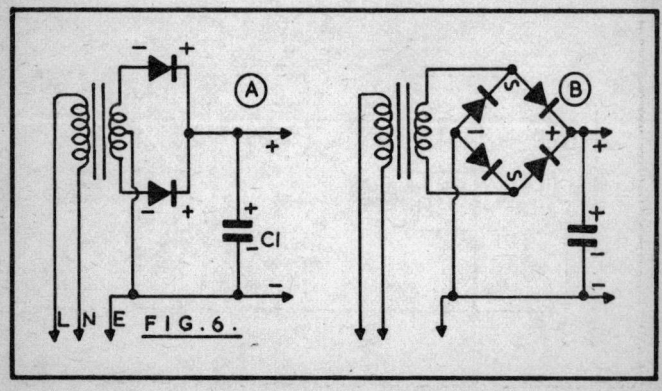

FIG.6.

(Figure 5) can no longer be supplied, control ceases and output voltage also falls.

C1 can usually be about 15V working, and at least 1,000 µF for each 1 ampere. If the secondary is of higher voltage, C1 can be rated to suit.

Circuit B is similar but uses an untapped secondary. A single 9V output replaces the 9—0—9V of circuit A. In each case the rectifiers may be 50PIV 1A, or 1N4001 and similar types, again up-rated where larger current is wanted. A bridge may be preferred for B, instead of individual rectifiers.

The circuits in Figure 6 are not intended to supply current directly, but to provide an input for regulators such as those in Figure 5 from which the required 5V will be obtained.

Figure 7 is a complete supply. Here, a 6.3V heater type secondary is conveniently used to provide some 7V to 8V or so across C1 (1000 µF – 2500 µF), according to load. The diode ZD1 maintains Tr1 base at approximately 5.6V, for approximately 5V output. C2 (47 µF) is for electronic smoothing, and is sometimes omitted. C3 (220 µF) provides a low impedance reservoir as abrupt changes in current drain arise with LEDs, decoders and other ICs.

In building this and similar circuits, draw current from a 3-pin plug with a 2A or other low rating fuse. This allows earthing of the secondary, for safety. A neon indicator (small

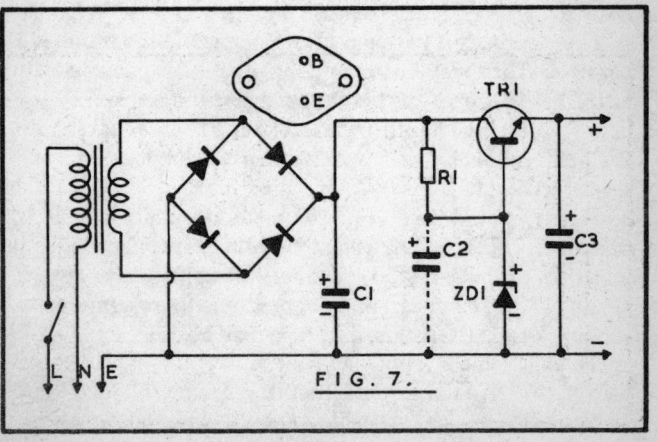

FIG. 7.

neon with 270k to 470k series resistor) may be placed across the transformer primary.

An earthed metal case is preferred, and Tr1 can be attached to a small heat sink, or may be mounted on the case. Except where the sink itself is fixed with insulated bushes and washers, Tr1 is isolated electrically from the sink (or case) by means of a TO66 insulation set for the 2N3054, or TO3 set for the 2N3055. This has a thin shaped insulator to go between transistor and sink, and bushes to allow fixing while maintaining insulation between transistor body and case or sink.

A tagboard or tagstrips will serve to mount the capacitors and other items. Positive and negative output sockets will allow easy connection of external equipment.

A check of output can be made under anticipated load conditions before first using the supply.

IC REGULATOR

The LM309 integrated circuit regulators are particularly useful for 7400 type ICs and other 5V supplies. These regulators automatically provide 5V. They have internal current limiting, and thermal shut down, thus protecting themselves and the power supply from a short circuit or severe overload. The LM309H has a TO5 case, to which a finned heat sink can be fitted, and is suitable for 200mA. The LM309K is of TO3 size, and suitable for 1A.

These regulators have only three connections, in, out and negative. They may be mounted on-board in some applications. Here, distribution of power is at a higher voltage, and each board with its ICs has its own regulator. This has advantages for large equipment, as it avoids extensive distribution of a regulated supply.

Figure 8 is the circuit of a complete 5V supply, able to provide in excess of 1A, using the LM309K regulator. SR may be four individual 1N4001 rectifiers for up to 1A, or any similar 1A 50PIV types, or 1.5A rectifiers for maximum rating. Alternatively, a bridge type may be used.

There is some latitude in T1 secondary, but approximately 8V should be found across C1, at least, under full load, Some 6.3V transformers are suitable; others will not allow this to be

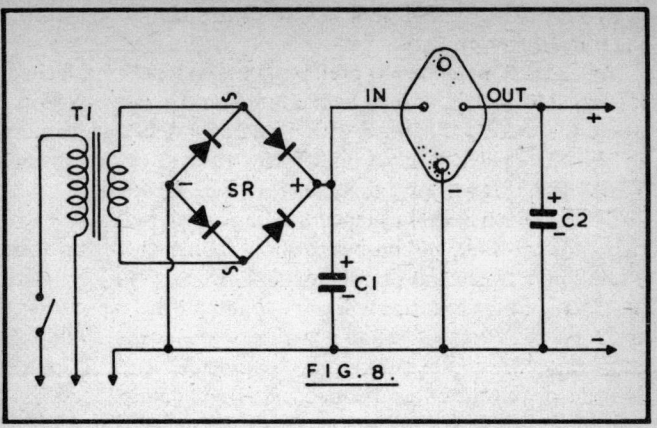

FIG. 8.

maintained. A 6.3V winding of fairly large current rating, such as 2A or more, is most likely to be successful. Otherwise T1 secondary can best be of slightly higher voltage rating, such as 9V. A centre-tapped secondary, A in Figure 6, can be used instead (typically 9—0—9V or similar). If the voltage across C1 falls below about 7V, regulation may be lost.

The LM309K is mounted on the chassis, panel or case, or an internal metal bracket, and this is common to the negative line. Heat to be carried away is not very great. As example, with 9V input, and 5V output, at 1 ampere, only 4 watts dissipation arises in the IC.

A meter check prior to first use should show that the out-out is very close to 5V. This is not related to the input voltage of the IC, except where T1 provides insufficient voltage. C1 may be 2200 μF – 4700 μF, 12V and C2 100 μF, 10V.

NIXIE SUPPLY

Around 250V DC is required for the Nixie, but only about 1mA to 3mA will be needed per numeral. Figure 9 is a suitable supply. The voltage across C1 will be approximately 1.4 times the RMS secondary voltage of transformer T1. C1 can be 350V, 8 μF to 32 μF, and R1 is a bleeder, and can be 220k 1 watt. The rectifier can be a 250V 40mA bridge, or four

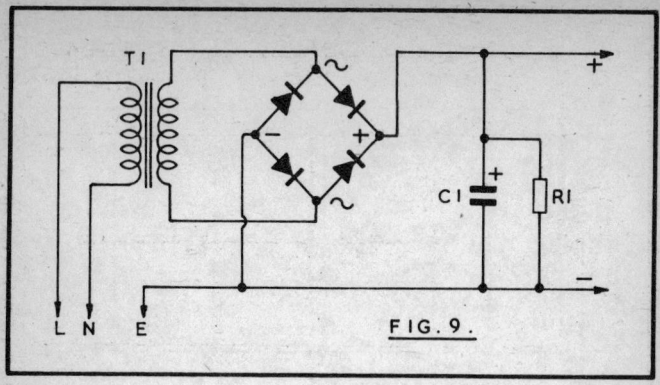

FIG. 9.

individual rectifiers if preferred. Half-wave rectification is also feasible.

Operation from 250V will be through an individual 33k resistor for each GN4 tube, or 47k for XN13 and 56k for XN3 tubes, with 220k to 470k for a decimal point. Low voltage reduces brilliance or results in no display or imperfect numerals. High voltage causes a reduction in tube life.

With a particular transformer at T1, voltage may be reduced by placing a resistor between secondary and rectifiers or between rectifier positive and C1 positive, and reducing the value of R1 so that some additional steady load is present.

Draw mains power from a 3-pin plug having a 2A fuse, and earth the secondary circuit at negative, as shown.

The Nixie supply and 5V supply for the ICs can be obtained from one transformer, using also a low voltage secondary, and IC or transistor regulator, as shown.

Multiplier for Nixie Supply

A high voltage supply for a Nixie numeral tube can be obtained, without the use of a transformer having a high tension secondary, by using a voltage quadrupler, Figure 10.

A transformer with 12V and 24V secondaries would allow the 12V tapping to be used with its own rectifier, Zener diode and series transistor, for a 5V supply. The 12V and 24V windings in series will provide 36V input for the quadrupler. If placing the windings in series provides a reduced voltage,

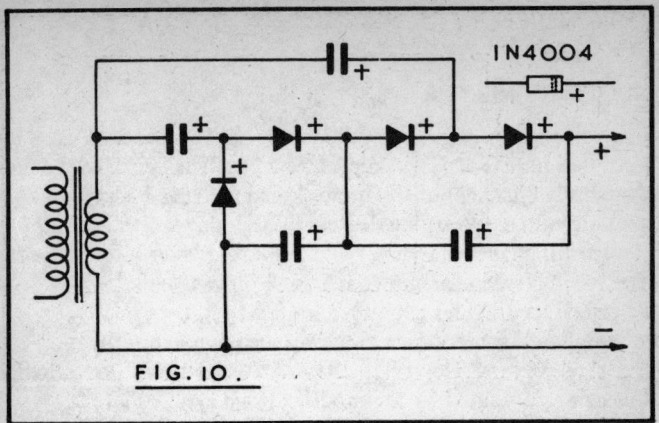

FIG. 10.

reverse connections to one.

Other transformers may of course be used. The eventual output voltage with no load is approximately 4 x 1.4 x RMS secondary voltage. Loaded, this falls rather rapidly, and was found to be 180V with the circuit shown.

Four 1N4004 or similar rectifiers are suitable. All the capacitors may be 4 μF, 150V. Capacitors of higher working voltage may be used, and all need not be of the same voltage rating. They should best be of about similar capacitance, and can be of larger value, such as 8 μF.

This circuit has no advantage over that shown earlier, except that a high voltage secondary is not required on the transformer. Rectifiers and capacitors are readily assembled on a tagboard.

PROJECTS

NIXIE NUMERATOR

Various popular games which combine skill and chance use a dice to obtain numbers. The simplest (e.g., such as Snakes & Ladders) are wholly chance, but the more skilled (such as Backgammon) are determined in varying degrees by the players' abilities, the element of chance allowing the less skilful player some success. The Nixie Numerator shown here will replace a dice. As described, it provides a number from 0 to 9, thus slightly 'opening out' some games. This is a welcomed modification for some players. However, where the usual 1 to 6 is required, this can be obtained by the modifications detailed later.

Figure 11 is the complete circuit (except for the power supplies). Two transistors operate as a multivibrator when the push switch is closed, values being chosen to give a visible yet quite rapid change of display. C1 provides electronic 'momentum', pulses continue to arise for a short time after the switch is released. This simulates the effect of many mechanical devices of similar kind.

Pulses go to 14 of IC1, the binary coded decimal decade counter integrated circuit. This IC provides outputs along its 1, 12, 9, 8 and 11 tags which are 0 to 9 in binary form. IC2 is a decoder-driver, and decodes its binary inputs at 3, 6, 7 and 4 into single ended outputs at pins 16, 15, etc., which connect to Nixie pins 13, 12, etc., to display numerals 1, 2 and so on.

It is apparent that as a random numeral display is wanted, the outputs 16, 15 and so on, to 2, could be connected to any of the Nixie pins 13, 12 along to 3. There would then be some chance, but permanently set sequence of numerals. This is readily provided by taking these ten leads at random to the pins. They are shown in correct order in Figure 11 because this does not effect the eventual working of the Numerator, and could cause confusion later, when referring to circuits in which numbers must appear correctly.

Thus the numbers 0–9 follow each other and repeat rapidly, so long as the push switch is held closed by the player, but stop a little while after this is released. The number obtained remains displayed until the switch is pressed again.

Tags 5 of the ICs are for positive supply lines, and 10 and 12 are common negative, as shown. Further details on building appear later.

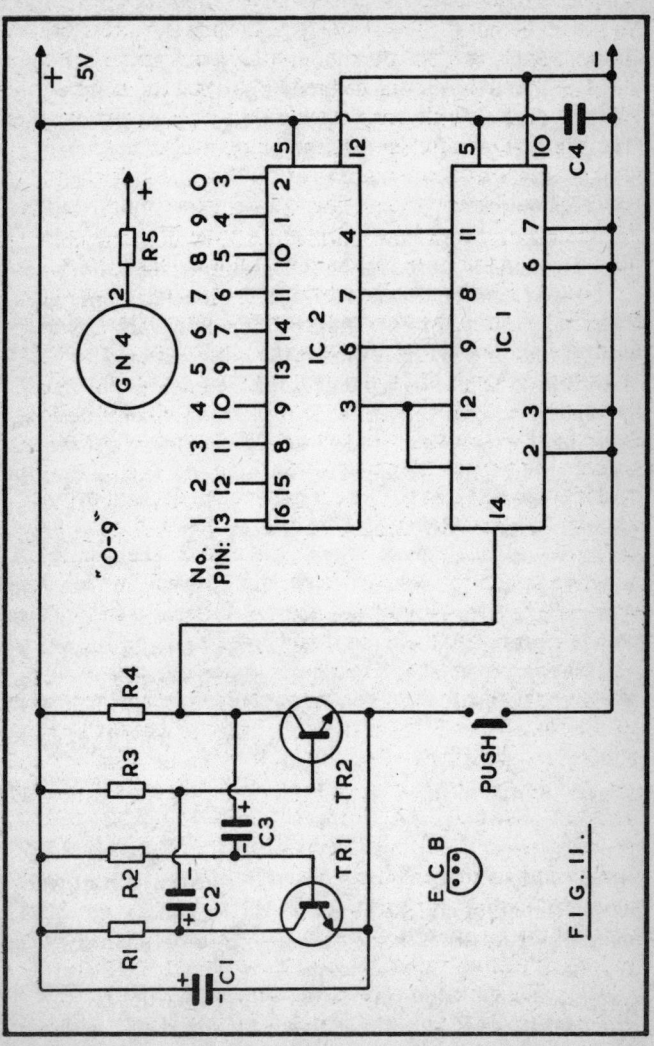

FIG. II.

Components for Nixie 0—9 Numerator (Figure 11)
   (Resistors ¼W 5%)

R1	2.2k	C4	47nF
R2	33k	Tr1/Tr2	2N3706
R3	15k	IC1	7490N
R4	1k	IC2	7441N
R5	33k ½W	Push switch	
C1	470 µF 6V	Nixie	GN4
C2	2 µF 6V	16 pin and 14 pin DIL holders	
C3	2 µF 6V	Board	

1—6 Numerator

The circuit of this is shown in Figure 12. A different integrated circuit is used in the IC1 position, and outputs are so arranged that the decoder-driver IC2 is operated only in such a way as to obtain 1 to 6. There is not a delay such as would result if the other outputs were merely disconnected.

In Figure 12 pins 6, 5, 4 and 3, for the unused numerals 7, 8, 9 and 0, are connected already if the Numerator is derived from that for 0—9, described earlier. If the tube is connected for 1—6 only, and haze appears, connect these leads also.

Components for 1—6 Numerator (Figure 12)
   (Resistors ¼W 5%)

R1	2.2k	Tr1/Tr2	2N3706
R2	33k	IC1	7492N
R3	15k	IC2	7441N
R4	1k	Push switch	
R5	33k ½W	Nixie	GN4
C1	2 µF 6V	16 pin and 14 pin DIL holders	
C2	2 µF 6V	Board	
C3	47nF		

Construction

Figures 11 and 12 are so similar that one general layout will be well suited for either. Thus only one is actually shown. The main difference arises in the connections to IC1, namely 9—6, 8—7 and 11—4 for the 0—9 Numerator, with 2 and 3 also grounded; while only 6 and 7 are grounded for the 1—6 Numerator, and IC1 to IC2 leads are 11—6 and 9—7 with 4 of

IC2 grounded. By using a wired board these changes are readily accommodated, or even the one type changed to the other.

Figure 13 shows wiring and layout for the 0–9 Numerator,

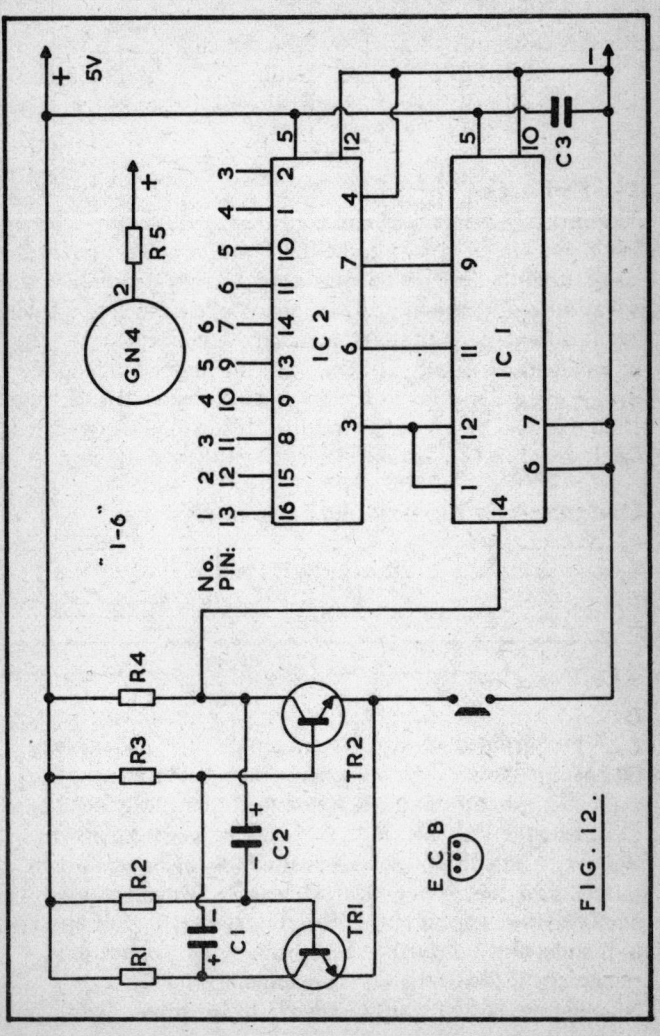

FIG. 12.

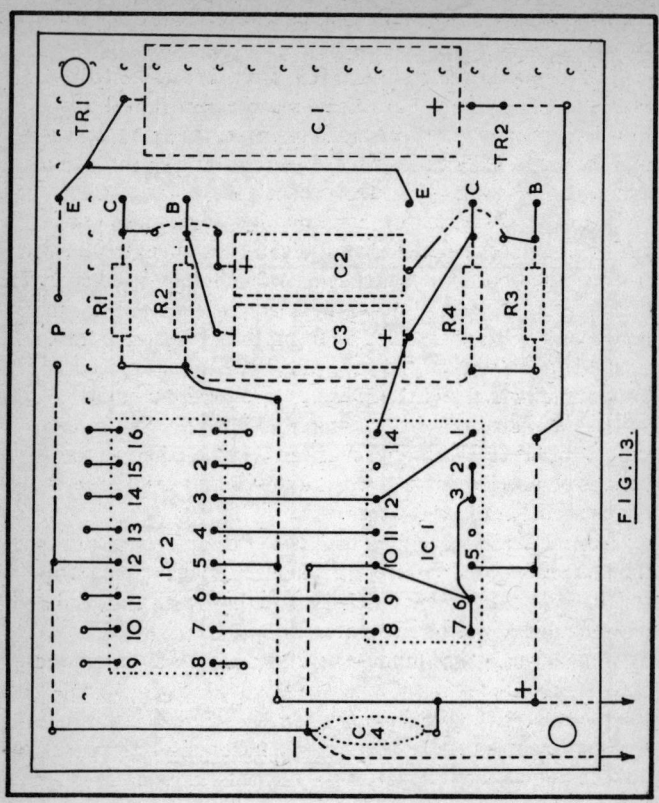

FIG. 13.

using a board 17 x 21 holes in size. Stout wire is not required, and is best avoided, 28swg being adequate. The underside of the board is shown. The ICs, components, and some connections, in broken lines, are on the other side of the board.

To take the IC holder pins, 0.1in matrix perforations are necessary. A small iron should be used for soldering, as there is not a great deal of free space where several connections or joints occupy adjacent holes. Resistor, transistor and capacitor leads come down through holes as shown, and connections can then be made to the various points.

Solder on red and black flexible leads for positive and

22

negative supplies. Also two flexible leads at P, for the push switch.

Ten external leads run from IC2, from 1, 2 and 8 to 16 (excluding negative at 12). These may be very thin flex; or thin single strand wire with small diameter insulated sleeving may be used. Take each of these leads in turn up through an adjacent hole, as shown, leaving them a few inches long.

Figure 14 shows a convenient method of mounting the GN4. The panel has an aperture slightly too small to allow the front of the tube to pass through. A 6ba bolt is fitted each side of this hole. A piece of insulating material, such as paxolin or hardboard, is cut and drilled to fit on these bolts, and this has a smaller hole, to clear the tube pins. The tube is placed between the panel and this piece, and set with numerals vertical, and nuts put on to clamp it in position. No force should be used. The flying leads can then be taken to the various pins, using short lengths of sleeving to form push-fit connections, in the way described.

This method of assembly can readily be extended for more than one tube, by drilling the panel and a long backing strip for the required number of digits. One bolt each end, and one between digits, will be adequate.

Where chassis type holders with fixing holes are used, these may be mounted on a sub-panel in the usual way, and will hold the tubes in place. Surplus circuit-board holders consisting of a ring of sockets may be mounted on a thin insulated panel, or may be put on the tubes after they have been fitted as in Figure 14. It is necessary to use thin material for the back strip.

It is also quite easy to mark an insulated panel with the pin positions, using ink on the pins, or carbon paper, and to drill small holes for them. Tubes can then be mounted by fitting them to the panel or board, and can be held in place by the leads themselves, fitted closely to the board.

Run a separate insulated lead away from pin 2, so that it can go to the series limiting resistor, which can be mounted on the board or on a tag strip.

Wiring as in Figure 14 will allow the panel with tube to be set flat over the circuit board and a depth of about 3in (76mm) inside the case will be sufficient. The push switch is on the

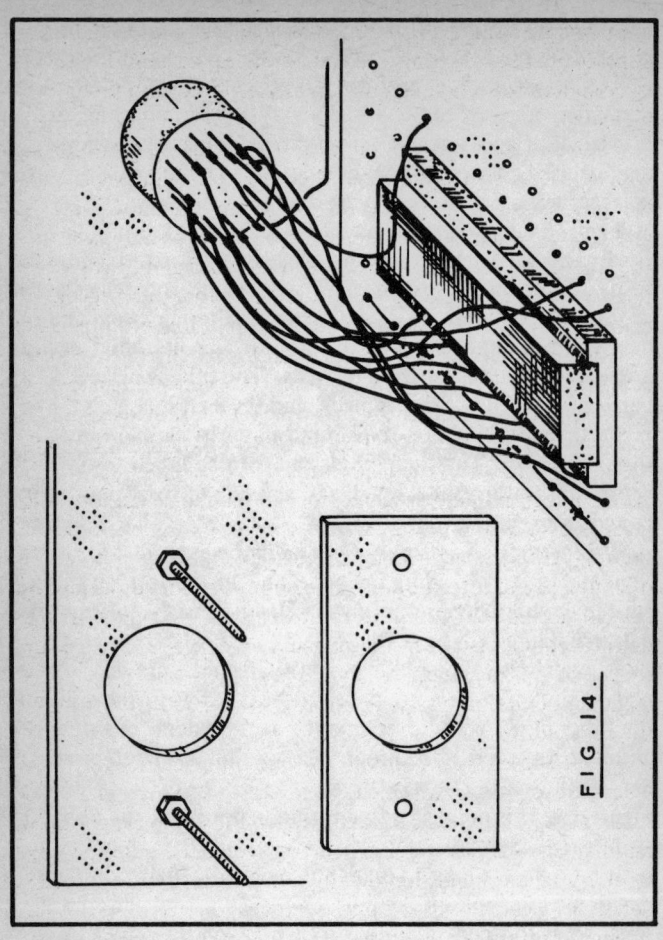

FIG. 14.

panel (or case top) near the tube, this giving easiest operation in play.

"STEADY HAND" WITH COUNTER

The amusing game which consists of threading a small ring along a stout wire or rod, while avoiding contact, can be improved by adding a counter. This is initially set at 0, but contact between the ring and rod advances this, and the test is to complete the whole 'course' in a given time with under 9 indicated.

Figure 15 is the circuit, and this includes a multivibrator which produces an audio tone when ring and wire touch, as warning. Suitable values for R1 to R4, and C1 and C2, are given, but there is actually considerable latitude here, and in the transistors used. Smaller capacitor values and lower value resistors raise pitch. C3 couples the audio to a small speaker, connected at S. This should be of 8 ohm or higher impedance.

The multivibrator part of the circuit is readily checked by itself, by completing the connection from VR1 to negative.

C4 charges each time ring and wire touch, moving 14 of IC2 negative, through R5. This advances the count, decoded by IC1, and taken to the Nixie tube.

Without this delay network, formed by the capacitors and including VR1, what is apparently a single contact between the ring and wire may in fact be revealed as a whole succession of contacts and interruptions, of extremely short duration, but able to advance the counter. In use, set VR1 so that a brief but definite contact between the ring and wire causes the speaker to sound, and the count to advance by 1. C4 tends to lengthen operation of the audio oscillator, and sound from the speaker.

It is feasible to use a longer and rather more difficult 'course' with this arrangement, compared to that where a single contact between ring and wire will lose the game. Book number BP48 Electronic Projects for Beginners, published by Bernard Babani (publishing) Ltd., contains further constructional details.

The Reset switch between 2 and 3 of IC2, and negative line, is normally closed. Momentarily opening this returns the tube indication to 0, for the next attempt. The Reset switch can be a push-button breaking circuit when pressed, or a spring-loaded toggle wired to give the same result.

Assuming that this game, like others, can be run from a general purpose AC operated power supply, no further switching is required. The transformer primary or main on-off switch will control both tone and IC supplies.

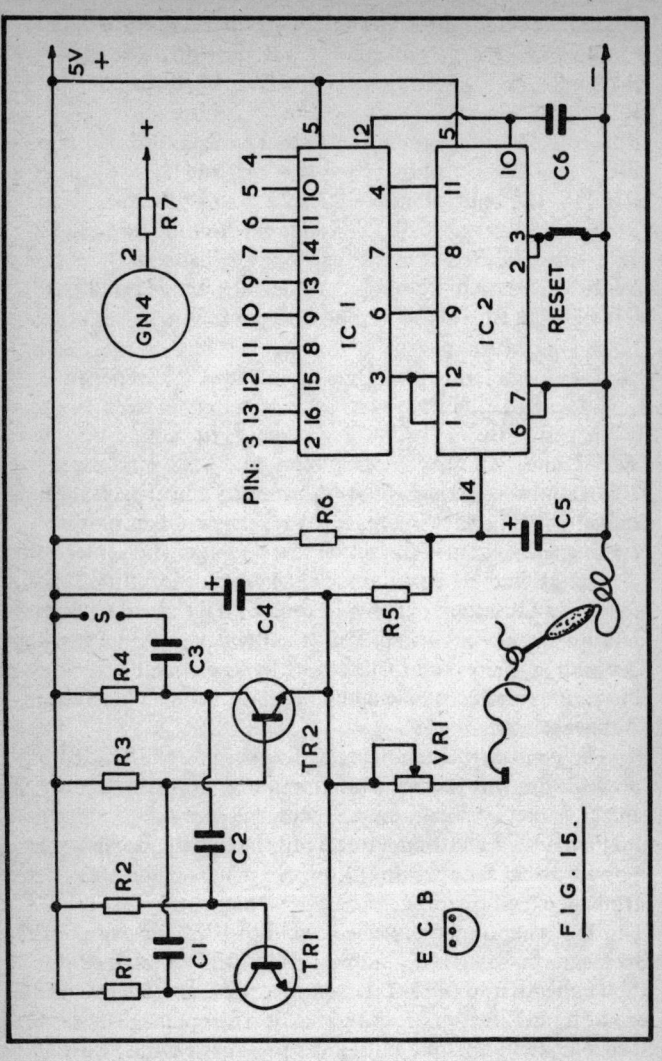

FIG. 15.

Components for "Steady Hand" with Counter (Figure 15)
(Resistors ¼W 5%)

R1	4.7k	C4, C5	10 µF 6V
R2	100k	C6	47nF
R3	22k	Tr1/Tr2	2N3704
R4	3.3k	IC1	7441N
R5	100 ohm	IC2	7490N
R6	1k	Reset switch	
R7	33k ½W	Nixie	GN4
VR1	250 ohm pre-set	14 and 16 pin DIL holders.	
C1, C2	0.1 µF	Board, etc.	
C3	0.22 µF		

"TWINKLE TREE"

This is an easy project for beginners, and has a number of applications. A sequence of ten light emitting diodes flash on one at a time, this being repeated so long as the circuit is operating.

The LEDs can be used on a small table decoration, in the form of a Christmas tree, evergreen or artificial, or a suitable picture; or they may be placed at various points on a hanging decoration, such as a bunch of mistletoe, or can be incorporated in other decorations or models. This provides an interesting and novel twinkling effect.

The same circuit can be used with the LEDs placed round a numbered board, so that they flash on in turn with a revolving effect, and can be halted by opening a switch.

Figure 16 is the complete circuit, including a small mains power supply for operation from AC mains. If running is to be from an existing supply, or batteries, omit ZD1, D1, R6, C4 and T1. The supply negative goes to the "E" line, and positive to tags 5. It should be approximately 5V, as explained.

The multivibrator Tr1/Tr2 produces pulses which form the input to the binary coded decimal decade counter IC1, at tag 14. This IC counts the pulses and provides a binary output at 1−12, 9, 8 and 11. Tag 5 is its positive line, and 10 negative line.

With counters having two or more numerals, as shown later, the IC can pass on a pulse to the "tens" section. Here, it is

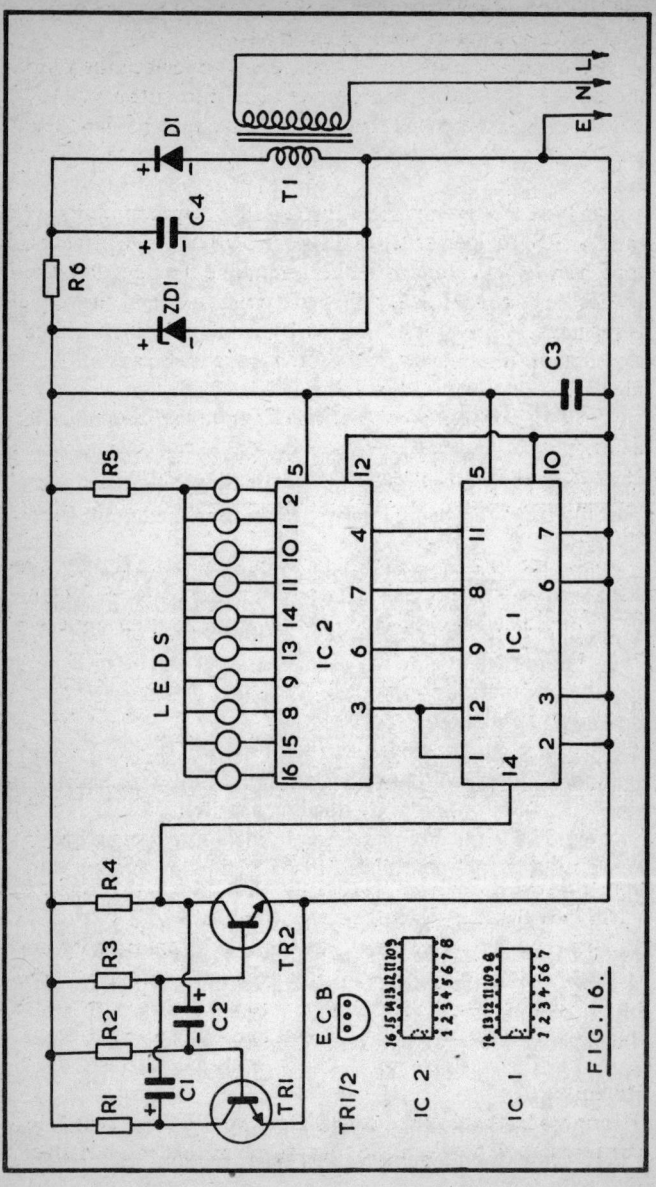

used alone and repeats the same series of outputs over and over so long as pulses are applied to 14.

IC2 is a decoder-driver. Its purpose is to receive the various inputs at 3, 6, 7 and 4, and decode them into outputs at 16, 15, 8, 9 and so on, along to 1 and 2. Thus each of these ten points (16 to 2) provides a circuit to the appropriate LED in turn.

All the LEDs are returned through the common limiting resistor R5, to operate from the same 5V line. The effect is, that each LED is illuminated in turn, along the line, and when the last is reached, this is followed by the first, and the sequence is repeated. Check the LED polarity, and mark this with red sleeving, or other means, if the LEDs have no means of identification (long lead or flat side).

For the 'Twinkle Tree' the ten LEDs are scattered at random, and any LED could be connected to any of the outputs 16 to 2. However, for the Roulette, where an apparent rotating indication is wanted, they should be in sequence in the order shown.

Only one LED is on at a time, so current drain is only some 50mA or so, and this makes battery running feasible. This can be 5V, obtained as explained (6V must not be used), or can be 4.5V from three 1.5V cells.

T1 (6.3V secondary) provides about 9V across C4, and R6 drops this for the 5.1V Zener diode ZD1. A meter placed across this diode should show about this voltage.

Mains current is drawn from a 3-pin plug with 2A fuse, and this provides safety earthing of the secondary and low voltage circuit. A 'double insulated' transformer specified as requiring no earthing may be operated without earthing. Naturally construction and wiring must assure that no mains voltages can reach the secondary or low voltage circuits.

With a battery or external power pack, assure polarity is correct. A thin red-black cord from the unit, with 2-pin non-reversible plug and matching socket on the power supply, will be convenient for taking 5V for this and other apparatus.

Roulette

The only change required is to place a push-button switch in the connection between the transistor emitters and negative line. The LEDs will then be illuminated in sequence so long as this is held closed, and one will remain lit when it is released. A slightly higher speed may be adopted, by changing R3 to 15k.

A board should have a circle drawn on it, and should be divided into 10 painted sectors, each with its own light emitting diode. The LEDs can be a push-fit in holes drilled in the board, or fit in grommets, with connections underneath. Join all positives together, and to R5. Ten thin flexible leads, running from 16 to 2, can then be soldered to the individual LEDs.

The push switch is best on top of the box. With battery running, place an on-off switch in one battery connection. The mains 'Twinkle Tree' or Roulette is intended to be unplugged from the mains when not in use so does not require a switch.

Circuit Board

Figure 17 shows components on top of the board, which is 0.1in matrix, 14 x 18 holes, with foils running vertically. Where foils are used as conductors, these are indicated with broken lines. It is necessary to make foil breaks between those parts of the circuit which must not be connected, under C1 and C2, between upper and lower rows of tags of each IC, between 1, 2 and 5 of IC2 and 14, 13 and 10 of IC1, as clear from Figure 17.

First place the IC holders in position, and solder the tags. Use a small iron and avoid excess solder. Horizontal wires in Figure 17 run on top of the board, and can be 24swg wire. Put sleeving on the lead from IC1 5 to R2, and also solder on red and black flexible wires for the supply.

Note that the base and collector leads of the transistors are differently positioned.

Foil breaks can be made with a few turns of a sharp drill, held in the fingers or in a handle. Examine these carefully to see that each conductor is completely cut, and that points of foil are not spread outwards to touch adjoining foils or joints.

If wished, the transistor multivibrator can be tested with headphones or similar means in advance, and should be heard to produce a continuous series of pulses at moderate speed.

The integrated circuits will usually have pins spread too

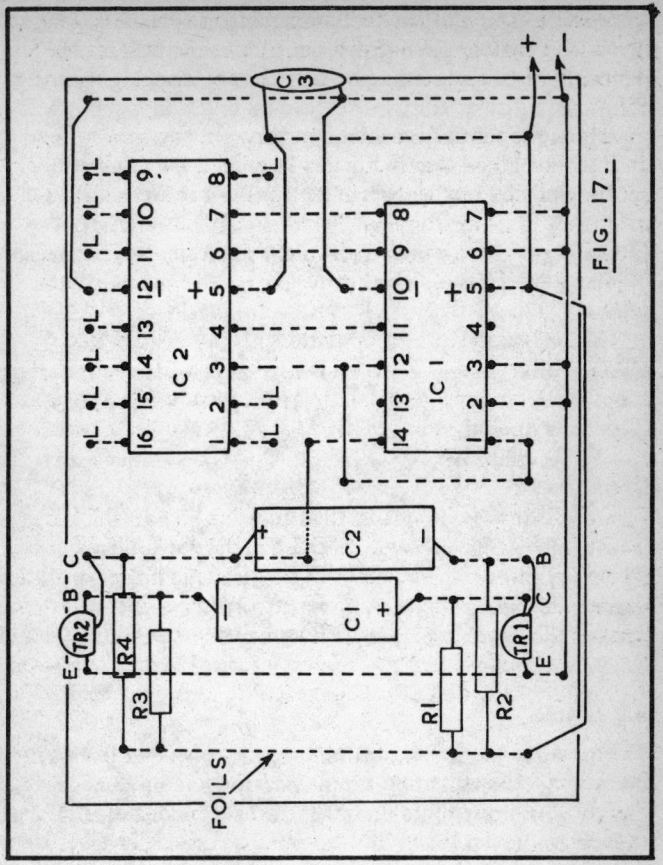

FIG. 17.

wide to fit the sockets easily. These can be bent in very slightly by pressing the IC on a flat surface. Assure the indentation or spot marking 1 is at the correct end of the socket, in each case.

The ten points marked L take thin flexible leads, soldered through the holes, and later connected to negative at the individual LEDs, as in Figure 16. The completed board can be fitted in a small plastic case, which may also hold the battery, where this is to be used.

For the Roulette game, disconnect the lead which runs

31

from emitters to 2 of IC1, and bring two leads from these points to a switch which makes contact when pressed. The periodicity is slightly increased (this may be done by reducing R2 or R3, or capacitor C1 or C2) so that motion is too rapid for anyone to select particular numbers.

It will be noted that in Figures 16 and 17 the numbering for IC1 and IC2 is when viewing these from above.

Components for "Twinkle Tree" and LED Roulette (Figure 16)
(Resistors ¼W 5%)

R1	2.2k	IC1	7490N
R2	33k	IC2	7441N
R3	33k	D1	1N4001
R4	1k	ZD1	5.1V 400mW Zener Diode
R5	220 ohm		
R6	33 ohm ½W	T1	Transformer with 6.3V secondary
C1	2 μF 6V		
C2	2 μF 6V	Push switch	
C3	47nF or 0.05 μF	DIL	14 pin holder
C4	1000 μF 12V	DIL	16 pin holder
Tr1	2N3706	10 off LEDs	
Tr2	2N3706	Board, etc.	

SIX-SPOT

The Six-Spot uses a combination of gates, operated from a divider which is pulsed by a multivibrator, and produces a display which resembles that of the usual six-sided dice. It can thus be employed for various games. The display changes continuously while a push-switch is held closed, and remains illuminated when the switch is released, so that the number can be read. The changing or running speed can be readily modified, but is a little too fast for anyone to select particular wanted numbers by releasing the switch.

The Six-Spot consists of three sections: the 2-transistor multivibrator, which produces pulses to drive the ICs; the four ICs with gates arranged to produce the required display, and the light emitting diodes, with their series limiting resistors.

Current required is about 60mA to 80mA or so, so that if

wished operation can be from a battery supply. An on-off switch is placed in one supply or battery lead, to switch off completely when the Six-Spot is not in use.

LED Section

The top of the case carries the LEDs providing the display, and also the push-switch. This is of conventional type, closing when pressed. This completes the negative circuit to both the multivibrator emitters, and the rate at which the display runs may be modified by altering the values of C1 and C2 here, or the values of R2 or R3, Figure 19. Smaller capacitor values raise the changing frequency.

To economise on space, the LEDs, with series resistors, are fitted to a neat top panel, and the multivibrator and ICs occupy a board to be mounted underneath.

Figure 18 shows the underside of the LED display. The LEDs are arranged to light in arrangements corresponding to the spots on a dice, as indicated. Only D comes on alone, indicating 1. More than one circuit may be completed. Thus A, B, F and G indicate 4, while with D also on, 5 is displayed.

R1 limits current through A and G in series. Similarly, R3 is for C and E, and R4 for B and F. However, R2 is for D only, so is of higher value.

Check the LEDs for polarity, with the supply and a limiting resistor, and if no polarity indication is present, provide this. A touch of red paint could be used, or short pieces of red sleeving. Do not bend the LED legs near the body, or the LED may fail to work afterwards.

A convenient size for the board is about 3 x 3in (76 x 76mm). The LEDs may be a push fit in holes drilled for the purpose, and be secured with adhesive, or can be a push fit in grommets which in turn can be put in the holes.

Thin flexible leads will run from the points shown, to 3, 4, 8 and 11 on the IC board. Also provide a red flexible lead for the common positive.

Components for LED Board (Figure 18)
R1, R3, R4, 120 ohm ¼W 5%
R2 270 ohm ¼W 5%   Panel approx. 3 x 3in.
7 off 7mm or similar LEDs. (76 x 76mm)

FIG. 18.

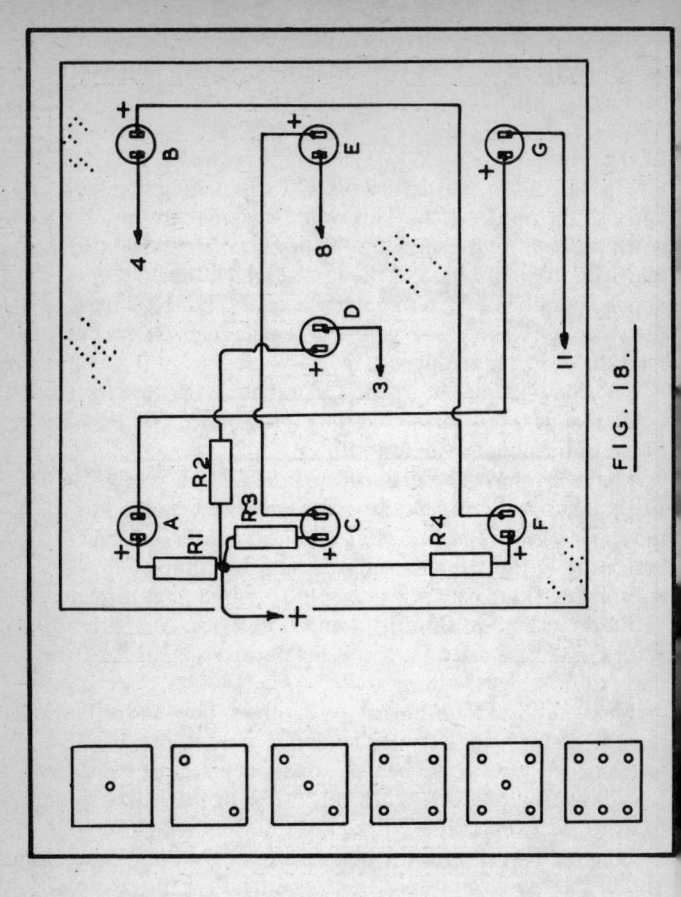

The LED board and LEDs can be tested by taking the positive wire to a 5V supply. Temporarily taking 3 to negative should light D. Similarly, 4 should light B and F, while 8 should light C and E. Connecting 11 to negative of the supply should light A and G. If any circuit fails to operate, check LED polarity in particular.

Multivibrator Section
This occupies the left of the board, Figure 19. The board is 0.1in matrix, 23 x 32 holes. Note that the emitter circuit runs

FIG. 19.

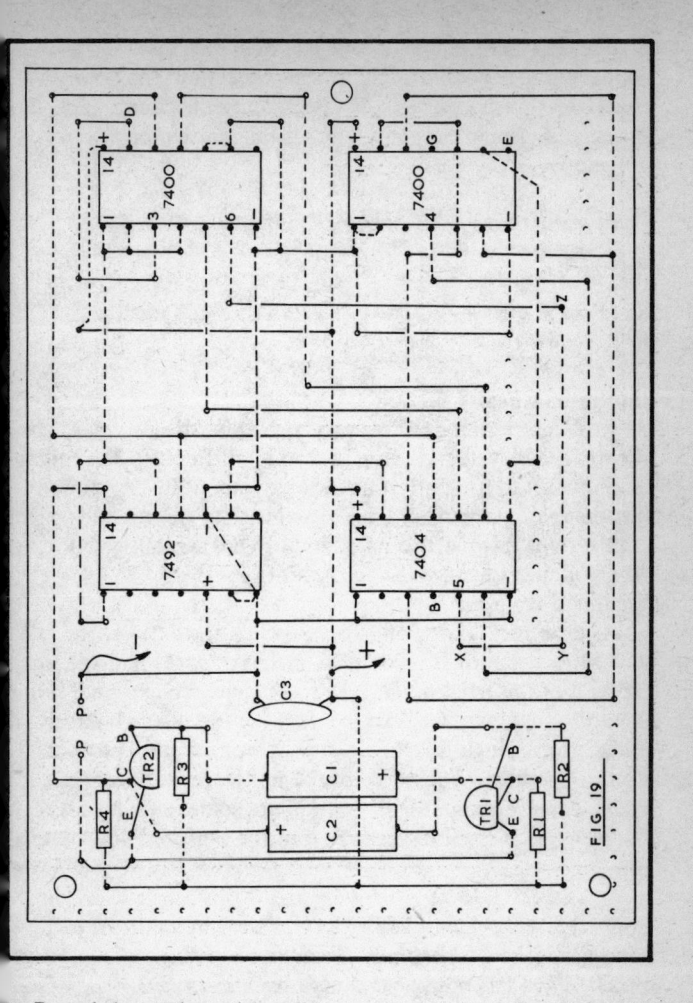

to P, and the push switch will be connected from P to P, so that it completes the negative line.

Other points are straightforward. Where C1 and C2 are not insulated, they should not touch each other.

With R1 to R4, C1 and C2, and Tr1 and Tr2 wired as shown, a 5V or similar supply can be connected to the resistor supply line, and emitters (latter negative). Pulses should then be heard

35

in high resistance phones across R4, or shown by a meter connected in series with R4. If this test is not satisfactory, check the transistors, resistor values, and other items here. The gating and display cannot operate if the multivibrator is inoperative.

Components for Multivibrator (Figure 19)
  (Resistors ¼W 5%)

R1 2.2k	C1, C2 2 μF 6V
R2, R3 33k	Tr1, Tr2 2N3706
R4 1k	

Integrated Circuits Section

At first sight wiring to these may appear a little complex, but if carried out systematically, it gives no difficulty. The method employed is to use fairly thin wire (say about 30swg) for all but the main positive and negative lines, and to run underneath connections from left to right, and top connections vertically.

Use holders for the ICs, and count pins from 1 to 14, viewing the board from the top.

As example, take the two bottom ICs, the 7404 at the left, and 7400 at the right. Cut a few inches of wire, and solder it to tag 5 of the 7404 holder. Draw the wire straight, and bring it up through hole X. Take it across the top of the board to Y and down through this hole. Draw it straight across to Z, and up through here. Take it vertically until level with 6 of the top 7400, down through the hole, to pin 6, solder and cut off. Similarly, 6 of the 7404 goes to 4 of the bottom 7400, and so on.

It is essential to use thin wire, and if this is bright tinned copper, it will solder readily. If kept straight and reasonably taut, no sleeving is generally required.

The 7492 has positive and negative at 5 and 10. The other ICs have positive and negative at 14 and 7, as marked. These circuits are made with slightly stouter wire, about 24swg is convenient.

Use thin flex or 30swg wire with 1mm sleeving for the leads from ICs to LEDs. Thus 4 of the 7404, marked B, is connected to LED B, 3 and 13 of the top 7400 connect to LED D, and so Take each lead through a hole in the board, to help anchor it.

Connections around the ICs may be ticked off against the following:

7492	7400
1, 12, to 1, 2, 4.	1, 2, 4 to 1, 12.
2, 3, 4, NC.	3, 13 to LED.
5 to positive.	5 to 10.
6, 7, 10 to negative.	6 to 5.
8, NC.	7, 9, 10 to negative.
9 to 13.	8, NC.
11 to 11, 12.	11 to 9.
13, NC.	12 to 11, 11.
14 pulse input.	13, 3, to LED
	14 to positive.

7404	7400
1, 7 to negative.	1, 2 to negative.
2, NC.	3, NC.
3 to 6, 12.	4 to 6.
4 to LED	5, 10, 13, to 12.
5 to 6	7 to negative.
6 to 4	8, to LED
8 to 9.	9 to 8.
9 to 11.	11, to LED.
10 to 5.	12, 6 to 3.
11 to 12.	14 to positive.
12 to 5.	
13 to 9.	
14 to positive.	

The ICs in this list occupy the same positions as on the board. NC denotes no connection.

When inserting the ICs, assure they are the correct way round and that all tags engage properly with their sockets.

The whole can be checked by applying power, and closing the switch connected to P–P. Indications should change until this is released, 1–6 then remaining, as the case may be.

A plastic, metal or wooden case is suitable, with accommodation for the battery supply, or with a twin cord for plugging into a 5V supply.

Components for Integrated Circuits Section (Figure 19)
7492
7404N
2 off 7400N
4 off 14-pin DIL holders
C3 4nF or 0.05µF disc ceramic capacitor
Push switch
On-Off switch
Board 23 x 32 holes, 0.1in matrix.

NOISELESS SWITCH

For testing, clock-setting and other purposes a manually operated switch allowing single pulses is useful. If an ordinary on-off switch is used alone, multiple intermittent contacts when making or breaking, almost instantaneous, may provide a series of pulses to which the counter or other device will respond.

The electronic noiseless switch in Figure 20 avoids this. It uses two of the four 2-input NAND gates available on the 7400 Quad NAND IC. The other gates may be left unused, or can be employed for other purposes, when the 7400 is incorporated in a clock, etc.

Each gate provides one input for the other, and the first pulse on moving the switch to P will step forward a binary coded decimal decade counter IC, such as the 7490, one place. Positive and negative go to positive and negative of the equipment (e.g. 5V supply). The lead from 3 is taken to the BCD IC, pin 14, or usual input.

The components may be assembled on a small board, with the 2-way switch fitted to the board, or to a panel or case. A spring-loaded switch can be used, or ordinary toggle or slide 2-way type.

Components for Noiseless Switch (Figure 20)
R1, R2   1k ¼W 5%
7400N
14 pin dual in line holder
Single pole 2-way switch
0.1in matrix board about 7 x 11 holes.

FIG. 20.

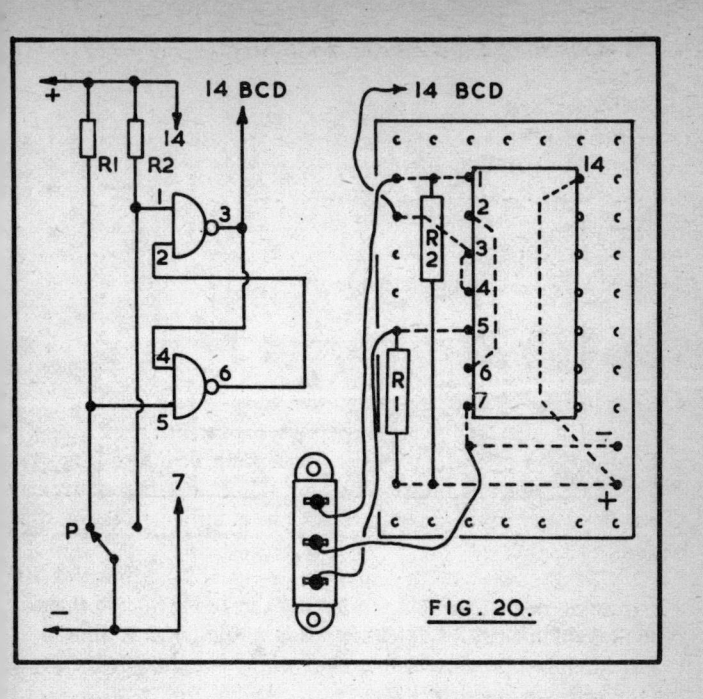

TESTING BCD DECODER-DRIVER

Figure 21 will provide a guide to trouble shooting a binary coded decimal decade counter and its associated decoder-driver, used to operate a Nixie numeral tube.

For testing purposes, a light emitting diode L can be used in series with R1 (about 390 ohm). A flexible lead with clip from R1 is attached to a positive (5V) point. A bright, stiff wire from the LED will serve as a prod, and the LED will light when this is touched on a negative point. LED polarity must be correct.

Logic outputs from the 7490 BCD will be high or low, along the 1–12, 9, 8 and 11 tags. These outputs, 1–12, 9, 8 and 11, are connected directly to the decoder 7441 inputs 3, 6, 7, 4. Therefore high-low tests made at the 1–12, 9, 8 and 11 points of the 7490 will agree with the

FIG. 21.

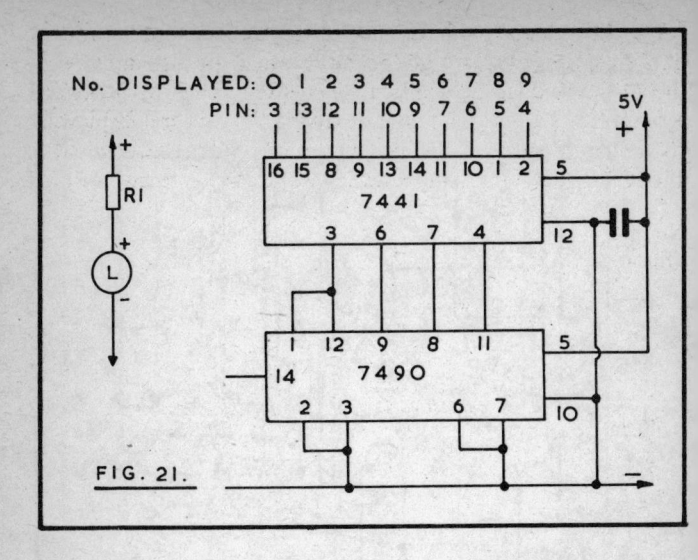

corresponding 3, 6, 7 and 4 points of the 7441, except when the conductor path is defective. This would indicate a fault in wiring, foils, or actual soldering, which can then be located.

Where the high-low states of the 7490 are followed throughout by the corresponding inputs of the 7441, connections from one IC to the other, here, are correct.

Where outputs from the 7441 (16, 15 ... 2) are connected to the GN4 Nixie pins 3, 13 ... 4 in the order shown, the numeral displayed will be 0–9 as shown, and will correspond to the binary appearing on the connections between the 7490 and 7441.

The reason for a faulty display can thus be sought as follows:

(a) **Display sequence wrong.** Check 7441 outputs and corresponding pin numbers.
(b) **Numeral(s) Missing.** Check Nixie as described. Check 7441 outputs, or substitute operating numeral, or check with HT current meter limiting current to 1mA. This will show if tube, connections, or 7441 faulty.

To check a 7441 is receiving correct inputs, the switch pulse

(Figure 20) can be used. Connect it to positive, negative and 14 of the 7490. Outputs from the 7490, which are also inputs to the 7441 as explained, can then be checked as below. Make one series of tests after each pulse, observing the numeral displayed. '0' indicates that the LED lights when its prod is touched on the IC tag listed, R1 being returned to positive as mentioned.

Number Shown	Tag 4	7	6	3
0	0	0	0	0
1	0	0	0	1
2	0	0	1	0
3	0	0	1	1
4	0	1	0	0
5	0	1	0	1
6	0	1	1	0
7	0	1	1	1
8	1	0	0	0
9	1	0	0	1

Logical 0 is actually around 0.2 to 0.5V at 10 to 13mA, and logical 1 may read from about 2.5V upwards with a meter only acting as load for the 7490. Logical 1 is typically 3.5V with a load drawing 0.8mA from the point tested.

If these results are not being produced by the 7490, it is either faulty, or not being used correctly (check for omitted leads, wrong voltage, tags not in holder sockets, etc).

If the 7490 produces these results, but the 7441 does not give the correct outputs, it is faulty, or circuit incorrect (check for shorts, omitted leads, etc.).

These tests should allow a fault to be located in a counter, driver, or display tube, or associated wiring, with a minimum of difficulty. Where more numerals are present, tests need apply only to that which is unsatisfactory. It may be easy to check some items by substitution, by unplugging them and placing them in a section which operates correctly.

The same type of test can be applied to circuits having LED numerals. Refer to the list of decoder-driver outputs by noting the sectors which will have to be illuminated for each number, as given earlier.

Other digital codes are also employed. That for the 7492 is as follows:

0	0000
1	0001
2	0010
3	0011
4	0100
5	0101
6	1000
7	1001
8	1010
9	1011
10	1100
11	1101

The "Grey" code changes only one digit or bit at a time, while the "Excess 3" code allows complementing by inversion:

0	0011	9	1100
1	0100	8	1011
2	0101	7	1010
3	0110	6	1001
4	0111	5	1000

LED RANDOM NUMERATOR

This is among the simpler devices to use employing a LED numeral, and the number displayed changes rapidly so long as a push switch is held closed, halting with a number shown when this is released. It can thus be used in all kinds of games, as a substitute for dice, or other methods of obtaining numbers for the players. Children may welcome the addition of zero and 7, 8 and 9, to the usual 1 to 6. If these are not wanted, they can be ignored when they arise.

A low voltage supply only is required. This can be 5V from one of the power supply units described, and current drain is 100mA to 150mA. The Numerator will also operate satisfactor from a 4.5V or 3-cell battery, and current drain is then about 80–120mA, fluctuating as the numeral changes.

Figure 22 is the complete circuit. Tr1 and Tr2 form a

FIG. 22.

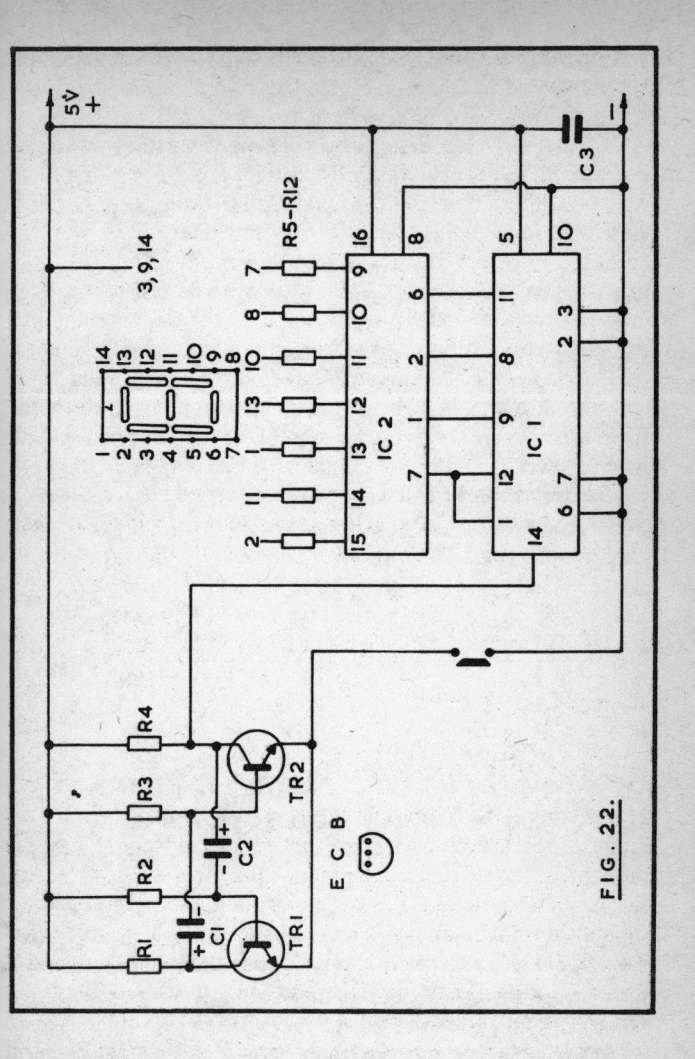

multivibrator, providing pulses as in other circuits described. Each time Tr2 conducts voltage drop across R4 takes pin 14 of IC1 negative, and when Tr2 is not conducting, and when the push switch is open, pin 14 is positive, via the limiting resistor R4.

IC1 is a binary coded decimal decade counter integrated circuit as used for the Nixie displays. Its outputs arise along 1, 12, 9, 8 and 11, and have been dealt with fully.

IC2 is a LED decoder-driver, which decodes these inputs, and provides outputs to light the correct segments of the 7-section LED numeral. Thus two vertical sectors are lit for 1, all sectors for 8, and so on, as explained earlier.

Resistors R5 to R12 are each in series with one of the 7 segments to limit current. The values of these resistors can be modified to some extent, to economise on current, or obtain full brightness with no loss of operating life for the numeral.

The numeral is a common anode type, and positive is shown connected to pins 3, 9 and 14. With some numerals, all these three positive pins may not be present, or may not be used. A mark shows the top of the numeral. It can plug in a DIL 14 pin holder, as used for IC1.

Components for LED Numerator (Figure 22)

R1	2.2k	2 x 2N3706
R2	33k	IC1 7490N
R3	33k	IC2 7447N
R4	1k	8mm common anode LED
R5 to R12 270 ohm each		Push switch
(see text)		0.1in matrix board
C1	2 μF 6V	16 x 28 holes
C2	2 μF 6V	2 x DIL 14 pin holders
C3	47nF or 0.05 μF	DIL 16 pin holder

Board Layout

Figure 23 shows layout on top of the board. Thin tinned copper wire will most easily solder and run through the holes from point to point. It is necessary to make small joints, using a lightweight iron. With 0.1in matrix board a large bit will tend to bridge two adjacent tags, and leave too much solder, while stout gauge wires can cause trouble particularly at the fragile holder pins.

The multivibrator section may be checked in advance, by taking a 4.5V or 5V supply to emitters (negative) and positive line. A rapid pulse should be heard at C2 positive, using high-resistance phones or similar means of checking.

FIG. 23.

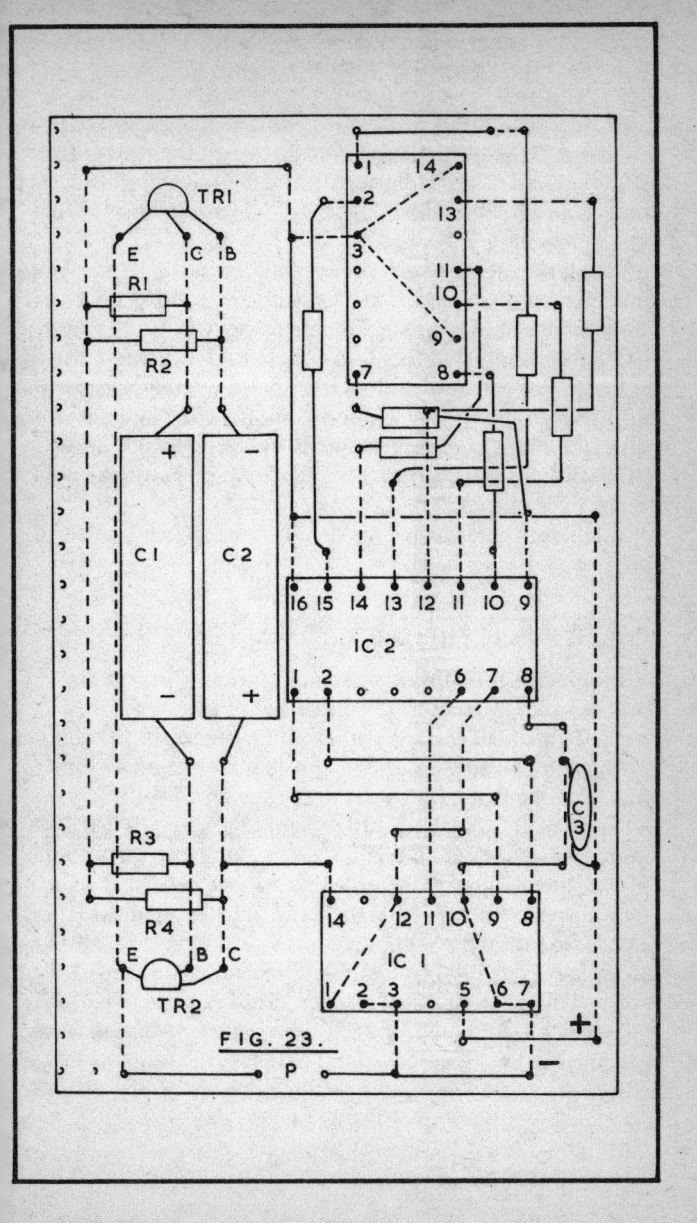

Resistors may be miniature 1/8 watt or similar components, though ¼ watt can be accommodated. Deal with the seven output tags of IC2 systematically, and no error need arise. E.g., 15 to 2 at numeral, 14 to 11, 13 to 1, 12 to 13, 11 to 10, 10 to 8, and 9 to 7. Shape the resistor ends to keep them away from each other, and if needed place sleeving on the wires in advance. Wrong connections will be shown by an incorrect display of some numerals or segments.

Not all the tags 3, 9 and 14 may be present, or be used for some numerals, as mentioned. The numeral and ICs must of course be put in their holders the right way.

Connect points P to the push-switch, best located on top of the case. This can be a shallow box, and can hold a battery if this is to be used. If so, include an on-off switch in one battery lead. Where the AC power supply is to be employed, fit red and black leads, with a non-reversible 2-pin plug, to take power from the 5V output socket.

Numbers should change rapidly until the switch is released, one then remaining displayed.

SCORER FOR BEZIQUE, ETC.

This forms the basis for other counting circuits for various purposes, so operation will be described in detail. Reference should be made to this section, for multi-numeral LED counters

When employed for Bezique, the counter is used with three numerals (one permanently set to 0, as required for this game) and an over-flow indicator LED. Counting is by 10s and 100s, as required, and after nine 10s, there is automatic carry to the 100s, in the usual way. After 990 is reached, a further score of 10 or higher lights the '1000 up' LED. This can indicate the game is won (two packs) or may be used to indicate 1000, to allow play up to 1990 (sometimes three packs are used). Addition arises in the usual manner. Thus if, as example, 390 had already been scored, and 250 were obtained for sequence, entering 2 at the 'hundreds' point advances the score to 590, while the 5 at the 'tens' point would move the display to 640.

FIG. 24.

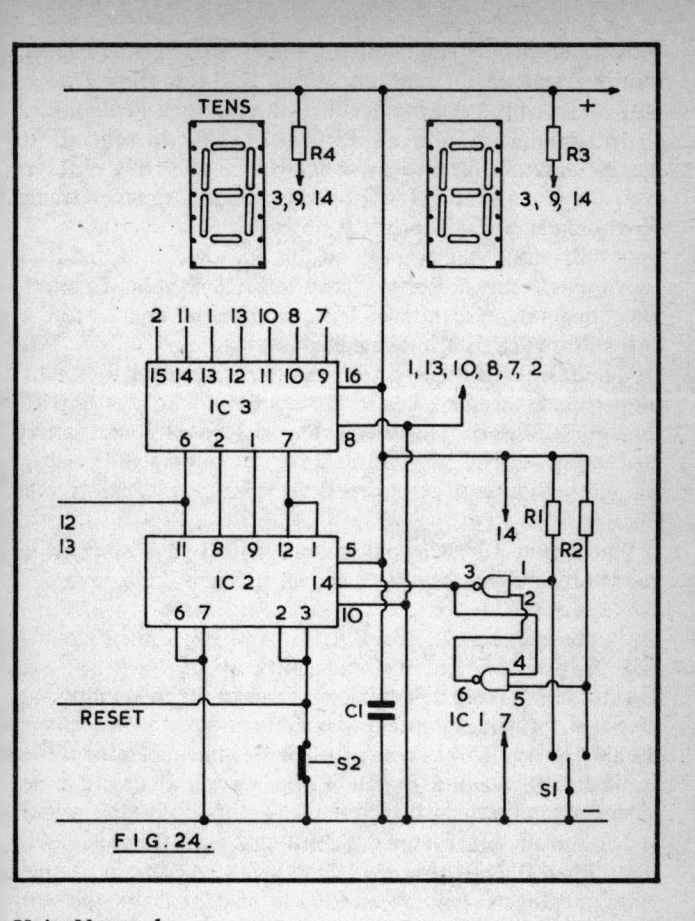

Units Numeral

As shown in Figure 24 this stays at 0. Six segments are wired to negative, and the resistor R3 is from common anode to positive, 5V supply.

Where a counter will require a changing numeral for units, this is operated from IC3 in the way shown for the 'Tens' numeral. E.g., IC2, IC3 and the numeral form the first section of the counter, and will show 0–9.

For the game scorer, scores advance by 10s, so no means of changing one numeral need be provided.

Noiseless Switch

The two gates of IC1, with R1, R2 and S1, form the noiseless electronic switch required to allow advancing the Tens numeral as necessary. Operation of this type of circuit is described earlier. A simple mechanical switch is very liable to give several contacts when working (not normally important) which would change the numeral incorrectly.

S1 is a spring loaded toggle switch, for preference, and is so wired that when released the Tens numeral changes. To alter the number by one, the switch moves from one contact to the other then back to its original position.

Thus for the Bezique scorer, to enter 20, press S1 twice, and so on for higher scores.

Output from the noiseless switch, at 3, goes to input of the divider, at 14.

Divider

IC2 divides by 10, giving pulses at 11, 8, 9, 12 and 1, which pass to IC3, where they are decoded, and light the correct sections of the numeral.

Positive and negative for IC2 are 5 and 10. These supply lines are often omitted in circuits, but must of course be provided in the actual equipment. C1 is to suppress switching transients, which can upset other circuits.

S2 is normally closed, and the counter then operates normally. S2 is a push-switch or other switch which can be momentarily opened. It then resets the numeral to 0, and also resets to 0 any other numerals controlled from the same reset line. When IC2 has operated IC3 in such a way that the numeral shows 9, the next pulse received at 14 of IC2 changes the numeral to 0 and also gives an output pulse at 11 of IC2.

Where a number of numerals are to be used one after another to count up to any required total, 11 will carry on to input 14 of the next section.

In this game scorer, 11 is taken to 12 and 13 of a further gate, in order that 100s may be added manually without interrupting the carry from IC2 when it arises.

Numeral

A slight simplification is introduced by using a common anode resistor R4, and wiring the segments directly to the decoder IC3. This allows seven direct connections from IC3 to the numeral. The value of R4 which is correct for two segments (number 1) is not of course ideal when more segments are drawing current through the same resistor. This results in a slight dimming, especially with 8.

In use, it has been found that with some types of 7-segment LED, the loss of brightness is not sufficiently noticeable to be in any way troublesome. But with LEDs of other manufacture, the loss is rather significant.

If the numeral is wired as shown, with one resistor, and the segments dim excessively when several are illuminated, then the resistor R4 should be removed. At the same time, place an individual resistor in each of the leads 15 to 2, 14 to 11 and so on, (see Figure 22). Alternatively, test the LED in the permanent zero holder first, connecting tags for various numerals, and noting the display brightness. If wished, the seven series resistors may be inserted originally, R4 being omitted. This also applies to other numerals in this and similar counters.

Hundreds

Figure 25 shows the circuit associated with this numeral. The carry over from 11 of IC2 passes to 12 and 13 of IC4. This inverter compensates for the presence of the following inverter, which receives input at 9, and provides output at 8, for 14 of IC5.

IC5 is the decimal divider, and IC6 operates the hundreds numeral in exactly the same manner as described for the tens numeral. The use of R7 alone, or fitting a resistor for each of the seven segments, applies here, in the way explained.

Input 10 of IC4 is from the noiseless switch operated by the push-button or toggle S3.

The four gates of IC4 are used here. Positive for IC4 is at 14, and negative at 7. Similarly, IC5 and IC6 have positive and negative applied to the tags shown.

The reset line carries on to 2 and 3 of IC5, and could run on to other dividers, if these were fitted for thousands and other numbers. Here, this circuit is associated with the 'Thousand Up' LED.

When the hundreds numeral has reached 9, the next pulse

FIG.25.

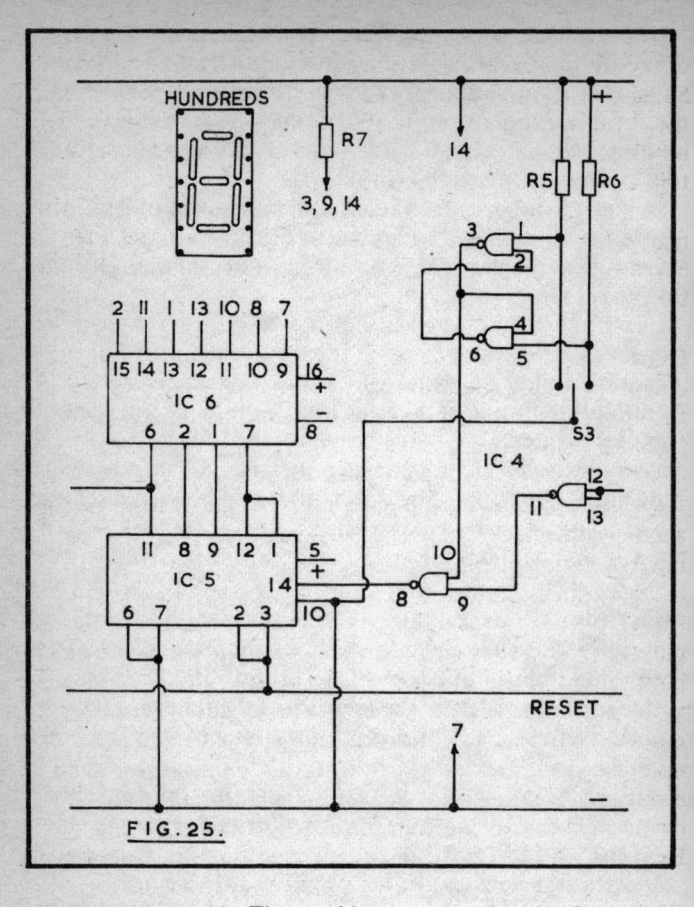

causes output at 11. This could run on to other dividers.

Figures 24 and 25 combined allow counting up to 990, for the game mentioned, with 10s by S1, 100s by S3, and resetting to zero by S2.

If the operating digits in these two circuits were used, the counter would read up to 99.

Where a numeral such as that for hundreds, Figure 25, is to follow one for 10s, Figure 24, and there is no need to be able to step forward the hundreds numeral individually, IC4 may be totally omitted. Then 11 of IC2 directly drives 14 of IC5.

50

If the manual advance provided by S3 were omitted, a score of say 100 would mean operating the 10s switch S1 ten times. S3 must be allowed to spring back to the positon which enables the pulse from IC2 to pass to IC5, or the hundreds count will not be automatic. It is thus best to use a spring loaded or toggle switch for S3, as described for S1.

A manual advance system similar to that furnished by IC4 may be used between minutes section and hours section of a clock, so that the display can be rapidly advanced to show the correct hour.

Overflow

An overflow indicator shows that the count provided for by the numerals has been reached or exceeded. Here, it is for '1000 up' or continuation to 1990 by playing with it lit.

The overflow has to ignore the positive going pulse, but responds to the negative pulse, and this is done with an inverter and small silicon controlled rectifier triggered by a capacitor input to its gate.

Figure 26 is the circuit of this part of the scorer. Output from 11 of IC5 swings positive at 8, and goes negative at the pulse after 9. The drop in collector current moves the capacitor positive, and thus triggers the SCR at its gate. The SCR remains in this condition until the cathode circuit is interrupted when resetting the numerals. This also takes place when initially resetting. As a result, the LED remains lit until another score is to be started from zero. (The operation of this type of device is covered in book number BP37 50 Projects Using Relays SCR's and TRIAC's by Bernard Babani (publishing) Ltd.).

Construction

Figure 27 is a suitable layout giving ample space, using a board 31 x 34 holes. Wiring around IC1 for the noiseless switch (duplicated at IC4) has been shown in Figure 20. Wiring for IC2, IC3 and this numeral can be seen from Figure 23 (refer to the previous note regarding series resistors). This is duplicated for IC5, IC6 and its numeral.

It is convenient to wire positive first, with thin red sleeving on leads, which can run along the top of the board above the LEDs. One negative, for 6, 7 and other returns, can run along

FIG. 26.

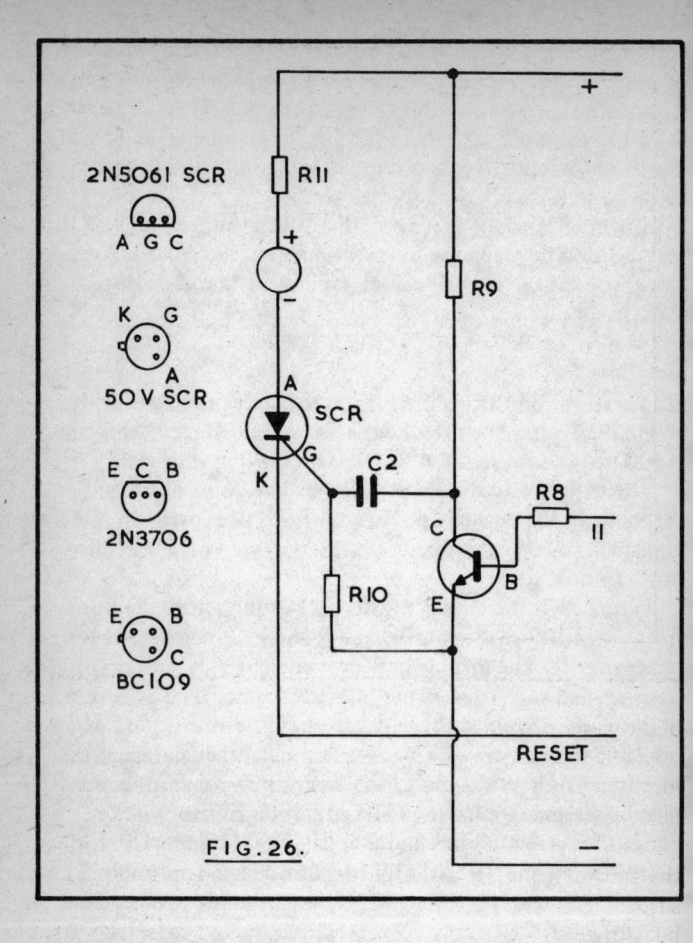

near the bottom of the board. Keep 2, 3 of IC2 and IC5 separate, for the reset line.

Bring out thin flexible leads for the switches, remembering that S1 and S3 must be connected in the manner explained.

Two units will be required, and shallow boxes can have twin leads to run from a common power supply. Switches should be so arranged that they are easily operated with one hand. The numerals can come behind individual apertures, or behind a single opening, covered with transparent material and having a

FIG. 27.

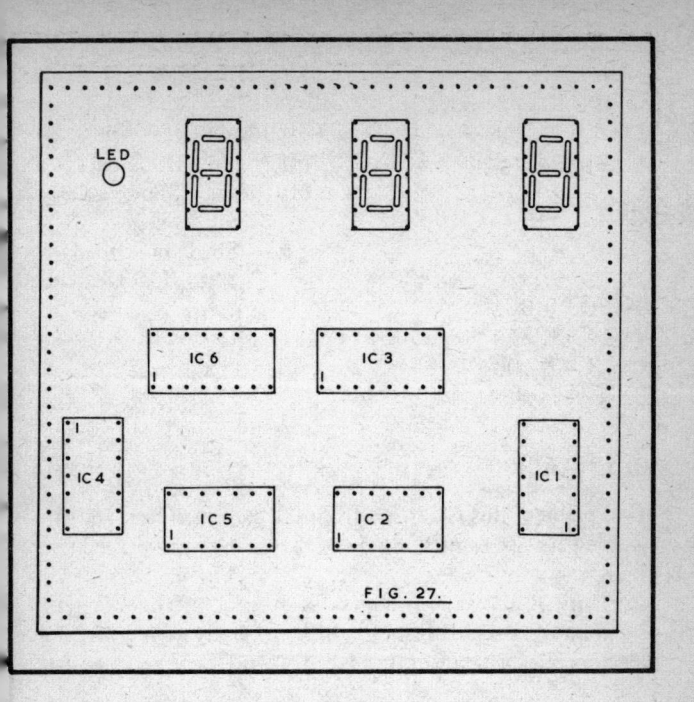

black paper mask to suit.

Faults are not very likely. If they arise, they can be easily localised. An incorrect zero will be from wrong connections here. Wrong tens displays, worked by S1, are most likely caused by errors between IC2 and IC3, or between IC3 and the LED numeral. If no advance arises with operation of S1, suspect IC1 circuits. Similarly, if the hundreds advance after 90, but not when S3 is used, suspect wiring around these two gates of IC4. If the count goes ahead correctly with 11 of IC2 taken directly to 14 of IC5, then IC4 circuits are suspected. Failure of the LED to light at 1000 after resetting and counting from zero would suggest an error in the transistor-SCR section.

Components for Scorer for Bezique, Etc. (Figures 24, 25, 26)
(Resistors ¼W 5%)

R1, R2, R5, R6 1k	IC3, IC6 7447N
R3 220 ohm	7 off 14 pin DIL holders
R4, R7 180 ohm	2 off 16 pin DIL holders
R8, R10 10k	3 off common anode LED numerals
R9 3.9k	
C1 47nF	S1, S3 Single pole change over spring loaded switches
C2 0.1 µF	
2N3706 or BC109	S2 Push to break switch
2N5061 SCR or 50V SCR	LED indicator
IC1, IC4 7400N	0.1in matrix board.
IC2, IC5 7490N	

MULTI-DIGIT COUNTER

As explained, this is derived from the earlier circuit, and is adapted for later units. Some of these will require a decimal point.

Figure 28 shows four digits, for up to 9999 (or 999.9, or 99.99 etc.). Each digit has its BCD or binary coded decimal decade counter, with outputs to its own DD or decoder driver,

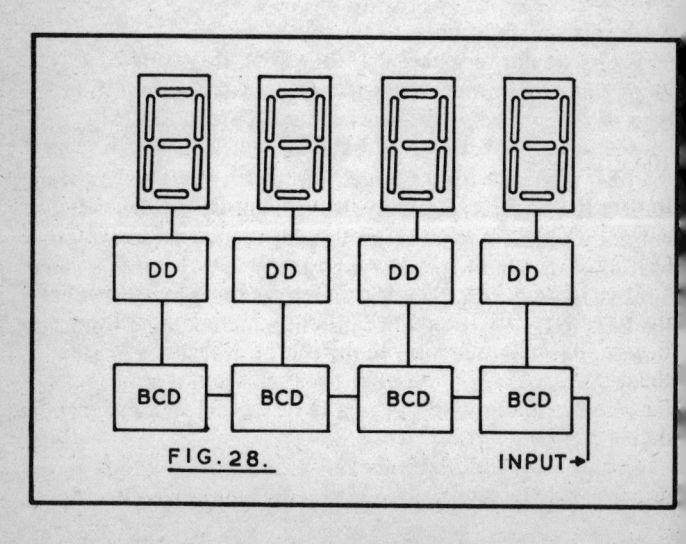

FIG. 28.

which in turn operates the 7-segment LED numeral.

Input is to 14 of the first BCD. Output from 11 of this BCD provides input to 14 of the tens BCD. In turn, this BCD provides output at 11 for input at 14 of the hundreds BCD, which from 11 similarly drives the final or thousands BCD at 14.

For a 2-digit counter, two BCDs, DDs and their numerals will be omitted. In the same way, omitting one BCD, the DD and numeral, gives a 3-digit counter. A 5-digit or 6-digit counter is made by adding one or two further BCD, DD and numeral circuits.

All positive and negative circuits are common, as shown earlier. Where numerals are to be reset to zero, the related BCD tags 2 and 3 are taken to negative through a switch which can be momentarily opened to achieve this. A common reset line and switch can control all the BCDs and their digits in this way.

For a decimal point, a 1k series limiting resistor can be used, to tag 6. Should a common anode resistor be used for the seven major segments, the value can be made up by placing a resistor of about 820 ohm between 6 and negative for the point only.

Multi-Digit Counter Components (Figure 28)
Each Digit: 7490N, 7447N, common anode LED, 14 pin & 16 pin DIL holders, 7 off 270 ohm resistors.
Also: Board (0.1in matrix), 47nF capacitor.

10X DIVIDERS

In some circuits it may be necessary to divide a count, without a display. Figure 29 shows two 10X dividers, providing 100X. One 7490 used in this way would provide 10X.

If one decimal divider is placed before the 4-figure counter, the latter will count up to 99,999 but with four significant figures. In the same way, two dividers as in Figure 29 extend the count to 999,999, but again with only four significant figures. The four significant figures may be adequate, depending on the purpose in view.

One or more additional dividers of this kind, without displays, may be switched into use with various circuits. As

FIG. 29.

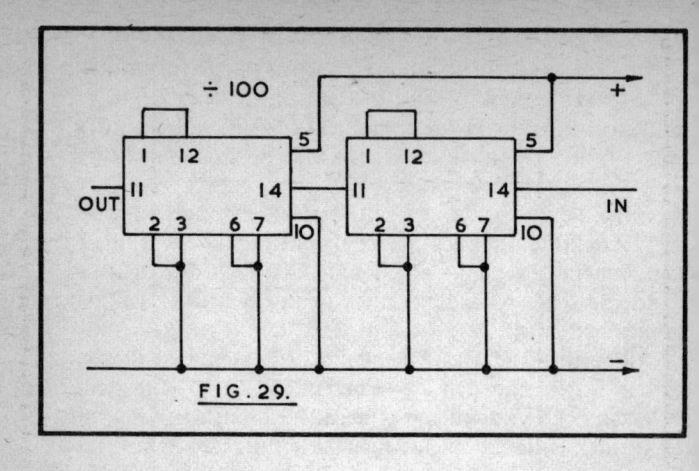

example, a process timer with three numerals could indicate up to 9.99 seconds; or up to 99.9 seconds; or up to 999 seconds. With the longer times, the absence of one-tenth or one-hundredth figures could be quite acceptable, depending on the purpose.

DIGITAL STOP-CLOCK

This stop-clock or stop-watch used with a three numeral display will give indications to one-tenth of a second up to 99.9 seconds, or to one second up to 999 seconds. There are two ranges, selected by a switch, according to the purpose and length of timing required.

The digital indicating section consists of a 3-digit counter, constructed as already described. Each digit will employ its own divider, driver, and numeral, and full details will be found in the circuits given. Two digits will give a useful display of up to 9.9 or 99 seconds.

The timing and range-selection circuit is shown in Figure 30. IC1 is the timer, with periodicity set by VR1. When S1 is closed pulses are obtained from 3 of IC1. S1 is the switch which is operated to time events and opened when the event ends, leaving the interval shown by the counter.

Pulses from IC1 are 10 per second. When the range switch

FIG. 30.

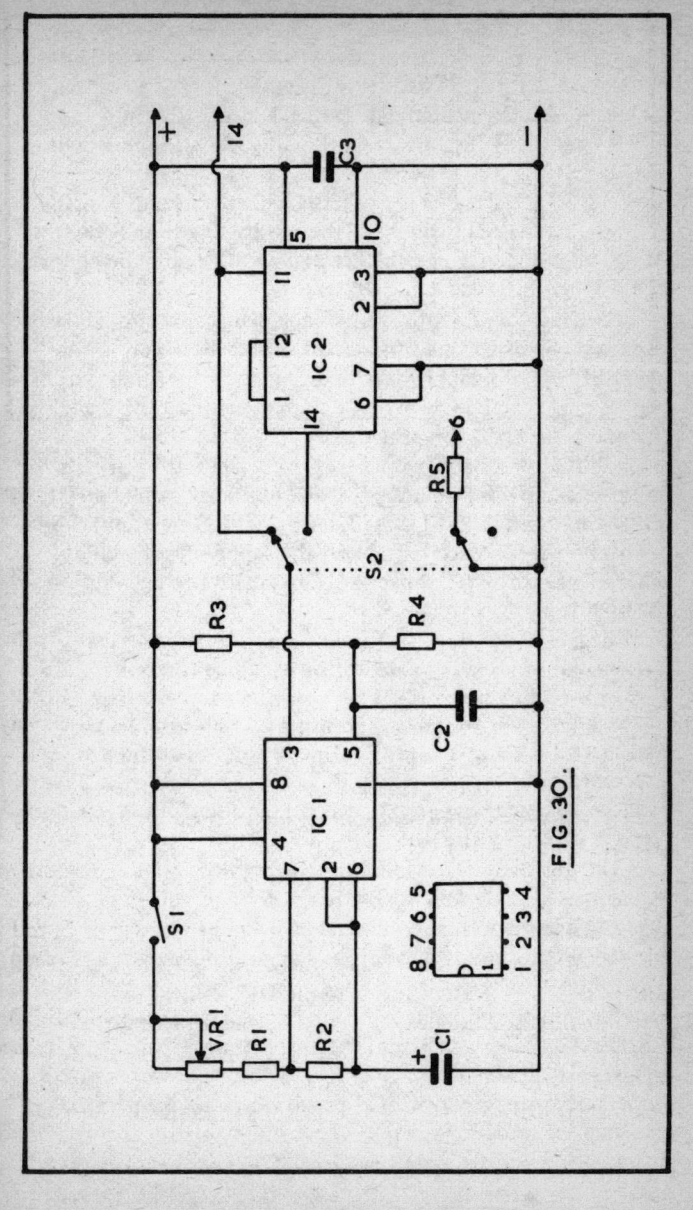

S2 is as shown, these go directly to the first divider of the counter, so that the first numeral operates at 0.1 second intervals. At the same time, R5 completes the circuit to Tag 6, so that the decimal point is illuminated. The Display will thus be up to 9.9 with the 2-digit display, or up to 99.9 with the 3-digit display.

When S2 is moved to the other range, pulses from 3 of IC1 go to the additional divider IC2, before passing to the counter. At the same time R5 and the decimal point cease to be connected. The display then reads up to 99 or 999.

The display with tenth-second included is appropriate for various competitive and other events. This, however, is not needed for photographic developing, and various other purposes. So the range can be selected as required, and best use made of the numerals which are provided.

With R1 at 47k and VR1 22k, the correct setting arose with about one-half VR1 in circuit. However, this depends on C1. VR1 is adjusted with S1 closed, until with IC2 in circuit the counter shows 60 each minute. If the supply is not well by-passed place a 100 μF capacitor from positive to negative on the board.

A degree of accuracy sufficient for many purposes can be obtained in this way. Where critical accuracy is required, the Quartz Stop-Clock described later may be adopted. The method in Figure 30 is easily accurate enough for numerous uses. C1 may be a 2.2 μF tantalum or 2 μF paper capacitor.

Components for Digital Stop-Clock (Figure 30)
(Resistors 5% ¼W)

R1	47k	C3	47nF
R2	2.2k	IC1	555
R3	100k	IC2	7490
R4	15k	8-pin DIL holder	
R5	1k	14-pin DIL holder	
VR1	22k or 25k linear pot	S1	On-off switch
		S2	2-pole 2-way switch
C1	2 μF 16V	Board	
C2	10nF	Counter as described	

Board Assembly

A board 14 x 15 holes will accommodate the components, Figure 31. Anchor points and cross-overs can be arranged by using holes and running leads above and below the board. Figure 31 shows the ICs from above.

S1 is mounted on the case, and also VR1, which can be adjusted with a screwdriver. For convenient timing of short intervals as accurately as possible, a twin socket may be fitted in parallel with S1, and a flexible lead with push-switch to hold in the hand can be plugged in here, S1 then being left open.

S2 is situated at any convenient position, and can be of the 2-way slide type. Take its central tag of one pole to 3 on IC1, and outer tags to 11 and 14 circuits of IC2. Solder R5 to the other pole, using the tag which illuminates the decimal point when IC2 is not in use. The first numeral of the counter then changes ten times a second. Illumination of the point shows when this range is in use.

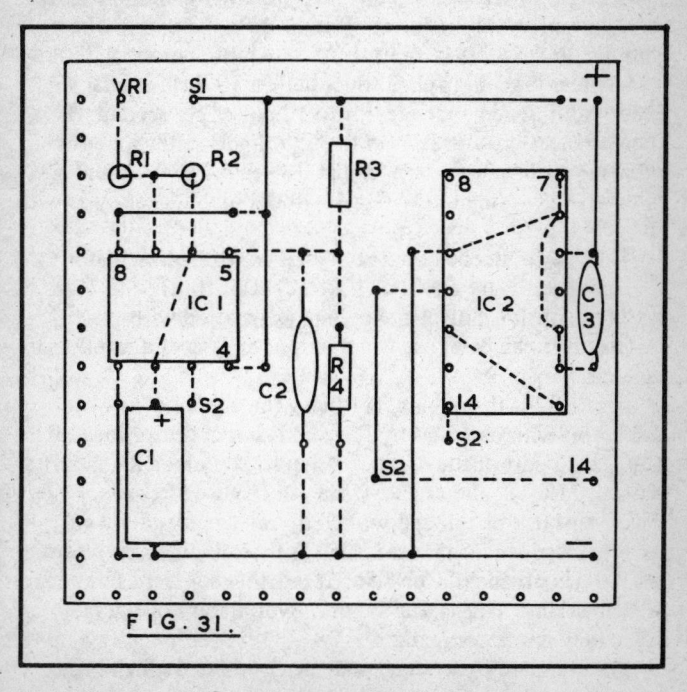

FIG. 31.

C1 should be a good quality component. Rotation of VR1 for best timing is least critical when most of the required resistance is made up by R1, and it is in order to use a lower value for VR1, and to increase R1 (or add other resistors in series) if required.

Take positive and negative to the common 5V supply lines of the counter, and a lead to 14 of the first divider in the counter.

QUARTZ STOP-CLOCK

For critical time keeping, the 555 adjustable timer can be replaced by a crystal controlled oscillator. This generally operates at a high frequency, and is followed by a number of dividers.

As example, it may run at 1MHz. That is, 1,000,000 cycles per second. Integrated circuits, each providing division by 10, would progressively reduce this to 100kHz, 10kHz, 1kHz or 1000Hz, 100Hz, 10Hz and 1Hz, or one pulse per second.

In the crystal controlled timer here, a 100kHz crystal is used. Four dividers reduce this to 10 pulses per second. This then replaces the pulses from the 555 timer, so that the first numeral of the counter can run at 0.1 second or 1 second intervals, according to the range selected by S2, exactly as with the 555 timer.

Thus the frequencies in the crystal controlled oscillator to replace the 555 are 100kHz, 10kHz, 1kHz, 100Hz and 10Hz, while the optional divider shown in association with the original 555 can be in use when output at 1 second intervals is wanted.

Figure 32 is the circuit, Tr1 being the crystal controlled oscillator, with base bias by R1, and R2 as collector load. C1 and C2 are part of the oscillator circuit. Trimmer T1, in series with the crystal, allows a very slight shifting of frequency. By this means it is possible, if wished, to set the crystal by an external frequency standard, such as that obtained by reception of 200kHz broadcasts, or 2.5MHz or other standard frequency transmissions. This is dealt with fully in details of the radio frequency harmonic marker.

The standard of accuracy may be considered much higher

FIG. 32.

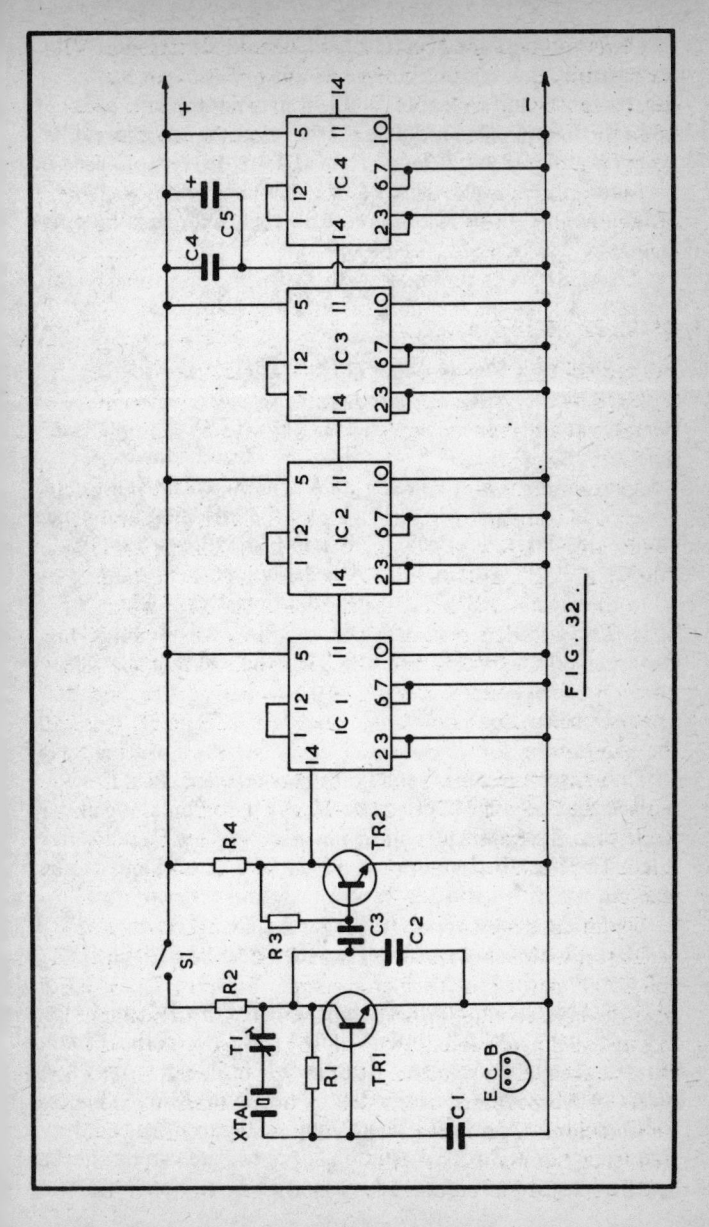

than necessary, with a crystal intended for 30pF series capacitance, and a 30pF fixed capacitor in place of T1. However, it is quite usual to fit T1 in such equipment, allowing fine adjustment. This is only a matter of a few hertz in 100kHz.

Tr2 is a buffer-amplifier. IC1 to IC4 are dividers, all used in the same way. Output from 11 of IC4 will go either to 14 of the counter, or 14 of IC2 in Figure 30 via S2, for the two ranges, as already described for the 555 timer.

Closing S1 starts timing, as with S1 for the 555 timer. IC2 of Figure 30 may be included on either board.

Components for Quartz Timer (Figure 32)
(Resistors 5% ¼W)

R1	1.5 megohm	Tr1, Tr2	2N3706
R2	2.2k	T1	60pF trimmer
R3	220k	XTAL	100kHz crystal for
R4	2.7k		30pF series resonance
C1	330pF silver mica	IC1 to IC4 7490	
C2	800pF silver mica	Crystal holder	
C3	10nF	4 off 14 pin DIL holders	
C4	0.1 μF	S1, board 40 x 16 holes, etc.	
C5	100 μF 6V	S2, IC2, R5 etc, Fig. 30	

Timer Construction

Components on top of the circuit board are shown in Figure 33. Drill holes to take the crystal holder sockets, and T1. The positive and negative lines can be 22 swg wire, run along under the board. Thinner wire will be more convenient for leads from this to the ICs. Connection are similar throughout, and can be checked against Figure 32. It may be found easier to do a particular circuit for all ICs at the same time. As example, 2, 3, 6, 7 and 10 to negative on all ICs. Then 1 to 12 with all ICs, and so on.

Solder on red and black flexible leads for the 5V supply lines at C5. Bring a similar lead up from 11 of IC4, to form the range selector switch connection. Thin flexible leads run from S1 points to this switch, described in reference to the 555 timer.

As mentioned, with T1 set to about half capacitance, a very high degree of accuracy should be obtained, and can be checked by observing the counter. Adjustment of T1 to obtain faster or

FIG. 33

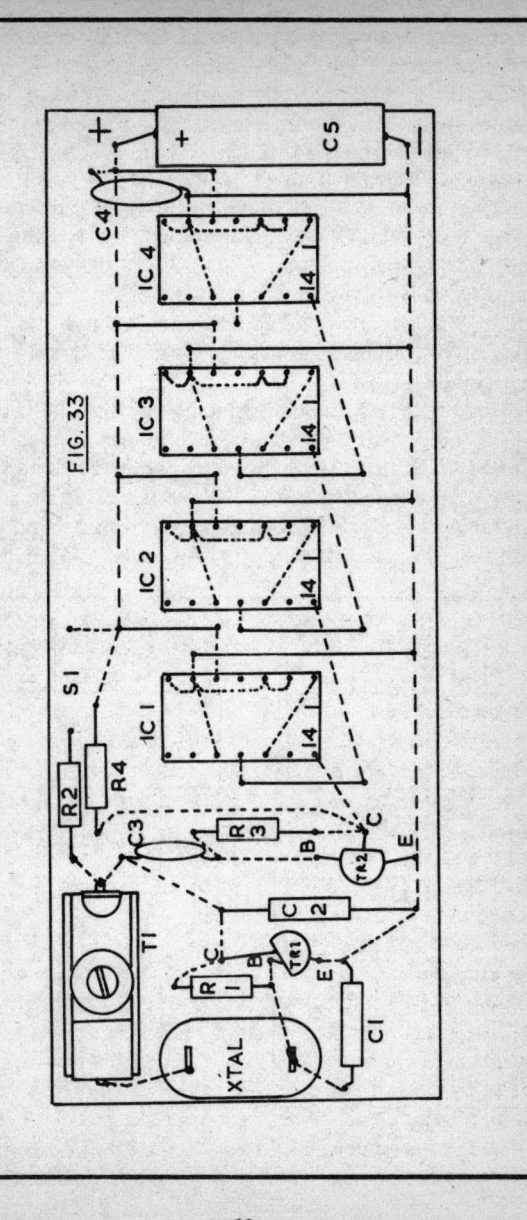

slower running has a very small effect, and can be carried out as explained for the harmonic marker for precision radio-frequency purposes.

Seconds-minutes Stop-clock
The numerals have been shown reading up to 99, or 999 with a 3-numeral display. This provides for the longest interval with a given number of digits, and is convenient for readings up to 9.9 seconds or 99.9 seconds. Where substantially longer intervals than 60 seconds will be timed, it is generally more convenient to modify the circuit to indicate seconds and minutes. Thus a reading of 90.5 seconds, as example, would read as 1 minute 30.5 seconds.

This requires two numerals running up to 60, before a pulse is passed to the next digit, and connections are shown in Figure 34. IC1 is the 7490, providing input for its decoder IC2, which is connected to the 7-segment LED (with series resistors) as indicated. This section runs from 0–9 normally.

IC3 is a 7492, and its output passes to the 7447 decoder IC4,

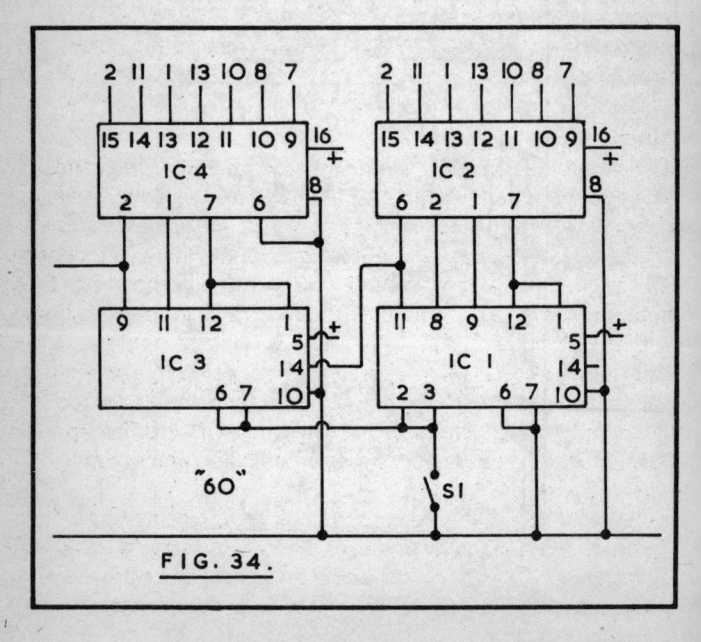

FIG. 34.

and to segments of the LED numeral (with limiting resistors). Connected in this way, IC3 and its numeral display up to 5, and when the numeral operated by IC2 receives a count after 9, the combined display of both numerals of 59 reverts to 00 and a pulse passes from 9 of IC3.

Input is to 14 of IC1. With pulses at 1 second intervals, IC1 and IC3 count up to 60 (00) in one minute, then providing the output pulse at 9 of IC3. This output pulse can pass to another similar counter, which will register up to 60 (00) minutes.

In order to indicate parts of a second, one divider with its numeral will be required before IC1 in Figure 34 for 0.1 second intervals; or two such dividers and numerals, for 0.01 second intervals. The way in which these are arranged and connected has been shown.

S1 is the reset switch, closed to count, returning the display to 00, and this operates at 6 and 7 of the 7492. Positive circuits are made throughout to the 5V line, in the usual manner.

Four numerals can provide readings up to 9 minutes 59.9 seconds, with five numerals giving up to 59 minutes 59.9 seconds or 9 minutes 59.99 seconds according to the range wanted.

Minutes-hours Stop-clock

The circuits shown allow timing up to 59 minutes 59 seconds. The next pulse will change the 59 second reading to 00, and pass a pulse to the 59 minute digits, which will then read 00. At the same time, a pulse is available from the minutes section. This may go to an overflow indicator, which will show that 1 hour has elapsed, and subsequent minutes and seconds readings will be additional to this.

It may be necessary to operate an hour counter, providing one pulse input for this each hour, from the minutes section. The simplest hours indicator is a single numeral, reading up to 9, thus allowing timing to 9 hours 59 minutes (plus seconds where fitted).

Alternatively, two numerals may be used, reading up to 24 hours. (Actual indication is to 23:59, returning to 00.00 at the hour). The circuit for this is shown in Figure 35.

IC1 and IC3 are 7490s. IC2 and IC4 are the decoder-

FIG. 35.

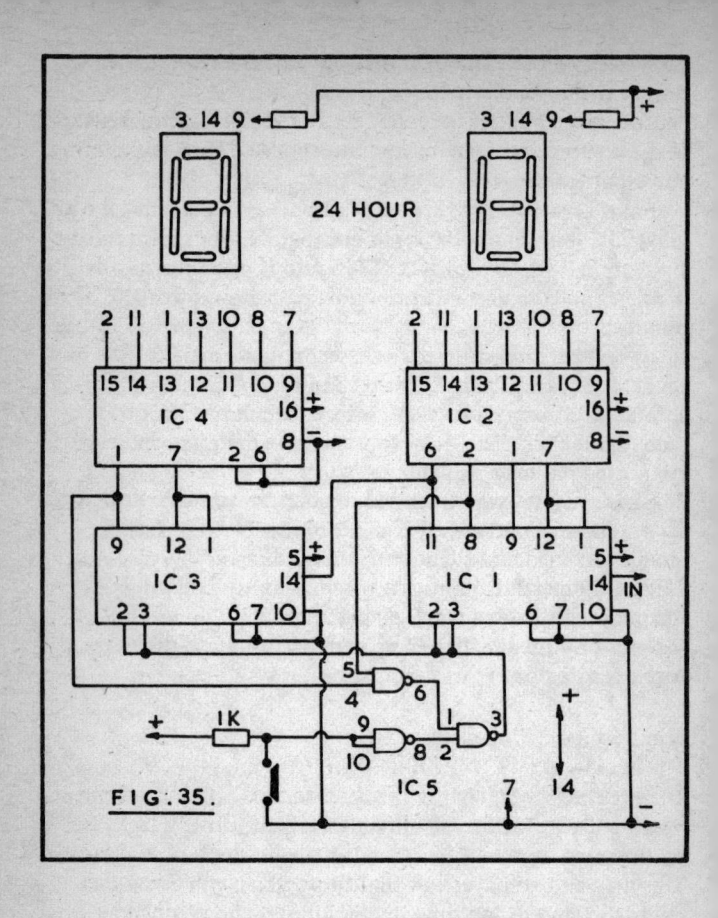

drivers for the 7-segment numerals. Counting must be so arranged that the units numeral runs to 9 when the tens numeral is at 0 and 1, but only to 3 with the tens numeral at 2, with the whole hours display returning to 00 at 24. Input to IC1 is 1 pulse per hour, from the minutes section, as described.

Three NAND gates in the 7400 IC5 are used to arrange the count, and allow resetting hours to 00 by the swtich. This is normally closed, so that 9 and 10 inputs are low, output at 8 thus being high. When the switch is opened, 9 and 10 go high, and 8 low, so 2 is low, and 3 high. The latter point controls both reset points of the ICs, thus opening the reset line and setting the numerals at 00.

The reset switch is necessary to set the hours when beginning to time an interval, and may be incorporated with the minutes and seconds reset switch, when fitted.

When hours are being counted, inputs at 4 and 5 allow 6 to go low and 3 high to reset automatically at 24 hours. This results from combining the outputs of IC1 and IC3 as shown.

Other parts of this section follow details given earlier. The numerals may have individual segment resistors, or run with a single resistor each, as explained, when this is satisfactory.

Thus the stop-clock (or clock) can have 4 numerals for hours and minutes, or 6 numerals to include seconds. Where 4 numerals are considered adequate for the timing purposes in view, the section to divide by 60 must of course be present, though no decoder-drivers and numerals are necessary. It will thus consist only of two ICs – the 7490 and 7492 with 1 second pulses in at 14, output from 11 of the 7490 to 14 of 7492, and output from 9 of the 7492 at 1 minute intervals for the minute section of the clock. Join 1 and 12 of each IC, as shown for the circuit which includes the decoder-drivers. Also provide positive and ground or negative as shown.

RESPONSE TIME INDICATOR

This will show a competitor's response time, in 0.01 second intervals, to 9.99 seconds. It is similar to a 'driving test' device which shows the delay between the presentation of a hazard, and the 'driver's' application of the footbrake. Numerous other tests can be arranged along similar lines, such as threading a needle, placing peas in a small bottle, turning cards or coins up and in sequence, fetching an object, or completing a short obstacle course.

Figure 36 shows the circuit. The counter is 3-digit, with decimal for 9.99, and runs from the crystal oscillator described, with division for one-hundred pulses per second input to the first numeral. An adjustable pulser could be used to simplify the circuit, its accuracy then depending on correct setting of this.

S1A is normally open, so that digits remain at zero. S1B is

FIG. 36.

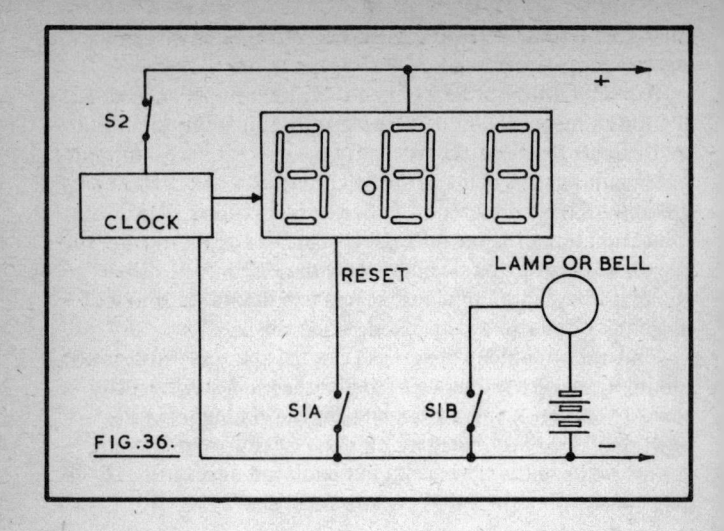

the second pole of this switch. The use of a second pole allows any lamp, buzzer or bell to be operated, from a supply of suitable voltage. Switch S2 is normally closed.

When S1A/B is unexpectedly closed by the 'tester' the indicating device operates, and the counter begins to register. The competitor must then open S2 as rapidly as possible, leaving the interval shown by the numerals.

A push switch, opening when pressed, is most suitable for S2, as this will then remain closed when released.

For puzzles, games or tasks taking longer, the timing can be extended to 99.9 seconds by fitting a 2-pole 2-way switch. One pole transfers the circuit to move the decimal point to the right, and the other introduces one further 7490 divider.

SOUND INITIATED TIMER

This may be used with any timer, but one showing minutes and seconds, with fractions, running from the crystal controlled oscillator, would be most suitable. Timing begins when a sound of enough intensity arises, and will normally be initiated by a blank firing starting pistol.

Timer circuits to operate the numerals have been shown.

Figure 37 is additional to this part of the equipment, and starts timing by automatically closing the reset line.

A small 60mm speaker of about 75 ohm impedance acts as the sound receptor, and is coupled to the first amplifier Tr1 by the capacitor C1. R1 provides base current here. Amplified signals arise across R2, and go to Tr2 via C2. Rectification by diode D1 moves the base of Tr2 negative. This is a PNP transistor, so collector current through VR1 rises, moving the gate of SCR1 positive. This triggers the silicon controlled rectifier into conduction, and it remains in this state so that the timer continues to operate.

VR1 allows adjustment of sensitivity, and this must not be too high, or other sounds can trigger the SCR and start the timer. Position the microphone (speaker unit) facing the firing location, with sensitivity to give reliable working.

The LED and series resistor act as an 'on' indicator and pass current to hold the SCR in conduction.

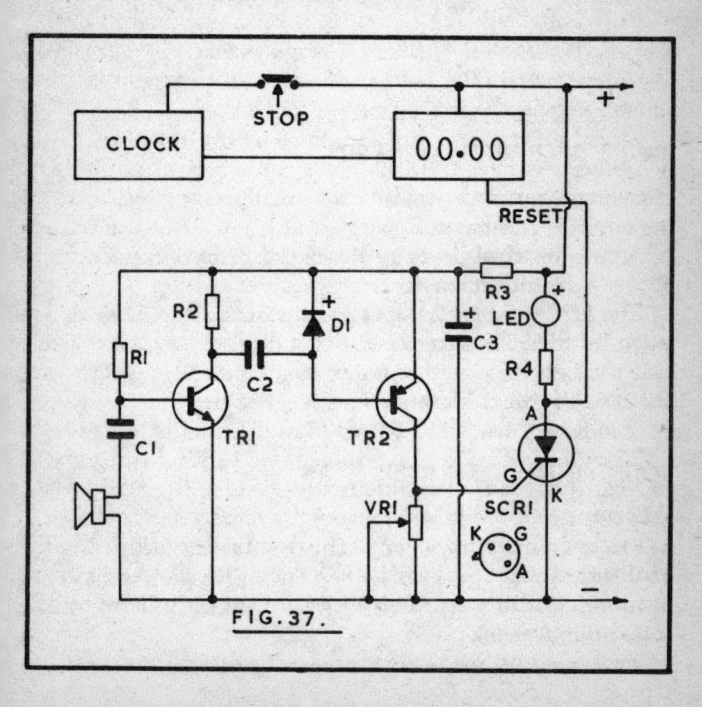

FIG. 37.

Halting of the timer can be by any of the methods shown, and not resulting in alterations to the count due to imperfect contact or other defects. Interruption to the counting chain is satisfactory provided it is at a point of quite high frequency. On the other hand, interruption by a slide switch, at the counter input, would usually result in a random jerk ahead of several figures, due to the imperfect contact described elsewhere. An electronic noiseless switch, such as shown, may also be adopted to stop the count with its display remaining illuminated.

Components for Sound Initiated Timer (Figure 37)
(Resistors 5% ¼W)

R1	1.8 megohm	C3	1000 μF
R2	4.7k	Tr1	2N3706
R3	270 ohm	Tr2	2N3702
R4	1k	D1	0A90 etc.
VR1	500 ohm (470 ohm suitable)	SCR1	50V 1A or similar type
C1	0.5 μF or 0.47 μF	2½in or similar
C2	0.5 μF or 0.47 μF	75—80 ohm speaker.

LIGHT OPERATED COUNTERS

A counter operated by interruption of a light beam has numerous applications. The circuit in Figure 38 is so arranged that the counter advances by 1 each time light is shielded from the light dependent resistor.

The LDR has a typical resistance of some 1k to 3k or so when lit, and 100k or more when not illuminated. When light falls on it and its resistance is low, it holds the base of Tr1 negative, so that collector current is negligible. The voltage drop in R2 is small, and input to 14 of the counter is high.

When the light path is interrupted, the LDR resistance rises to a high value, and base current via R1 and VR1 cause collector current to flow through R2, so that point 14 moves negative, advancing the count. The count is not further advanced by restoring light to the LDR. VR1 allows adjustment of sensitivity, and R1 is to avoid damage to Tr1 by careless adjustment.

In Figure 39 the circuit is so arranged that the counter

FIG. 38.

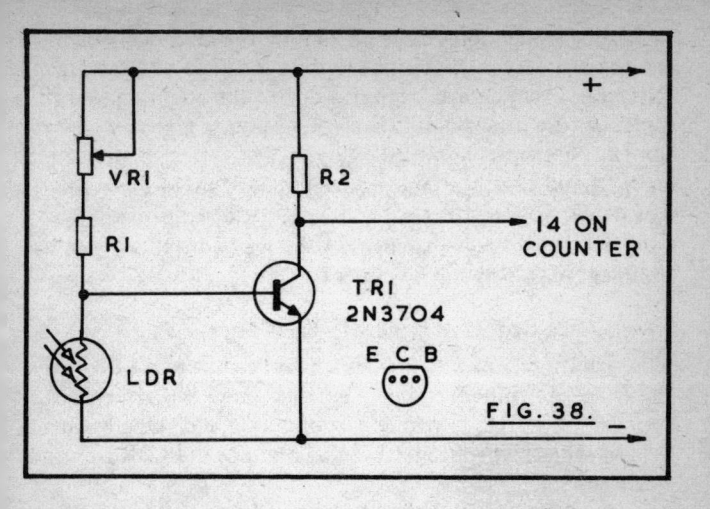

advances 1 each time the LDR is illuminated. Here, when the LDR is not lit, its resistance is high, so VR1 maintains the base negative. When light falls on the LDR, its resistance falls, providing base current, so that voltage drop in R1 moves 14 of the counter negative. Note that in this case the counter advances when light reaches the LDR, but the count does not

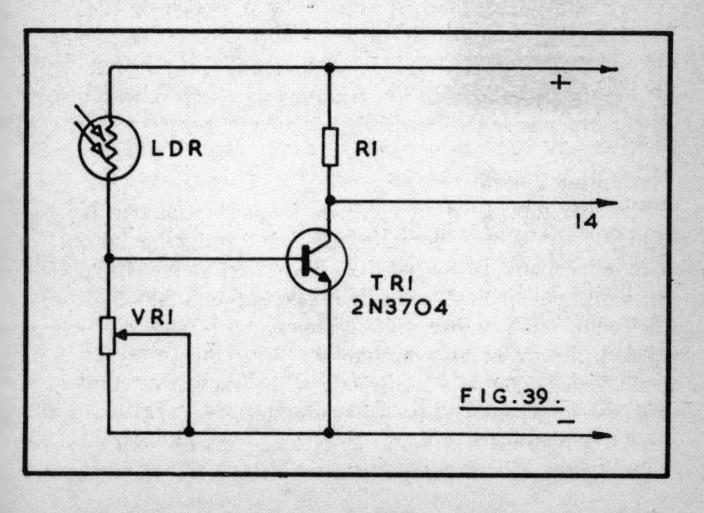

FIG. 39.

advance when illumination fails.

For these circuits the ORP12 or similar LDR will be suitable, with 2N3704, 2N3706 or BC109 for Tr1. In Figure 38 R1 is 33k and VR1 1 megohm linear, with 1.8k for R2. In Figure 39 VR1 is a 470k linear potentiometer, and R1 1.8k.

In each case a high resistance voltmeter may be clipped from Tr1 collector to negative line, and VR1 adjusted until suitable high and low readings are obtained with the illumination which will be present.

Persons Counter

This will show how many times a light beam across a doorway has been interrupted, and may be used where this gives access to a demonstration, separate part of a shop, etc. Operation is least critical where general illumination of the LDR is fairly subdued.

Fit the LDR in a tube, with the counter, one side the doorway, and directly opposite have a small, low-powered spotlight, or similar means of illumination directed at the LDR. With VR1 suitably adjusted, the counter will advance by 1 every time an opaque object comes between light source and LDR. Avoid any setting which results in the counter responding to flickering of the light at mains frequency.

A 3-numeral decimal type counter will register up to 999, and a manual reset switch will allow returning this to zero as necessary. Placing one decimal divider, without numeral, before the counter will allow counting up to 9990, with display to the previous ten.

Revolution Counter

Direct interruption of the light beam by a rotating part is sometimes possible. The beam may pass through the holes of a lathe faceplate, spoke spaces of a flywheel, or something of this kind. A small light on a flexible cord, with tube to direct the beam, will be convenient. Needless to say, care is exercised to keep clear of moving machinery.

The final count will be divided by the number of interruptions per revolution — as example, six, for a 6-spoked wheel. Where the test will often be made on one machine, or with one similar set of conditions, an extra IC (to divide by 6, or as required) may be placed between Tr1 and the counter.

For RPM in the simplest manner, employ a watch with seconds hand, and a noiseless switching system, operated by a push button, which can be held down for 1 minute. This gives easily enough accuracy for most such purposes. It also allows expansion of the reading, by timing, as example, for 30 seconds only, and doubling the count to obtain RPM.

For automatic counting, a timing gate is used. This opens for one minute (or generally such shorter interval as may be selected) and thus the counter registers for a known period, from which the RPM may be derived. Such a method becomes essential for very brief timing intervals, and will be found in the section on frequency counters.

Where light cannot be directed through a rotating wheel, the same result may be obtained by reflection from an area of white paint on the rim of the wheel, or some other convenient point. The light must then be directed only at this, and the LDR is shielded by a tube so that only reflected light is picked up. If the whole surface reflects, paint an area matt black.

6-Digit Frequency Counter

A frequency counter can best be assembled as a number of units, which may be tested separately. This avoids having completed the instrument, and possibly having some obscure fault which could be difficult to locate. It may be divided into the following section, as a matter of convenience in construction.

6-Digit Counter. This will allow counting up to 999,999, and with a central decimal point and kHz operation, allows reading audio frequencies and radio frequencies over a useful range. Frequencies will be shown directly. Thus 000.050 will be 50 hertz, 000.862 is 862 hertz, 001.500 is 1500 hertz, or 1.5kHz, and so on, up to full capacity. Frequency coverage can be extended by moving the point or changing the range by a multiplier, for frequencies higher than 999.9kHz. A LED numeral counter is indicated, but where a counter has been made with Nixie or other numerals, this will operate in the same way.

Clock. This controls the gating interval. Thus, if the interval is 1 second, pulses are counted for this period. As example, if the counter section registers 013.000, during 1 second, the frequency is 13kHz, or 13,000 hertz or cycles per second. Other ranges may be obtained by changing the gating interval. If this were reduced to one-tenth of a second, only 1300 pulses would be counted. So the range has in effect been multiplied by 10. If the gate is open for longer or shorter intervals, the number of pulses counted will be greater or smaller. A correct interval is thus necessary, and this is obtained by using a crystal controlled oscillator or clock.

Gate & Reset. The gate is opened and closed by a dual J—K flip-flop, so arranged that one pulse triggers it open, and the next pulse triggers it closed. The input signal, which is to be counted, can thus pass to the counter section for 1 second.

The counter has to be reset to zero after each display, or the counts would add one to another. This is arranged by opening the reset line then closing it before the count begins.

These methods of working provide a resolution which will normally be accurate to 1 digit.

Amplifier and Shaping. To allow various inputs to be dealt with an input amplifier is usually incorporated. This raises the signal levels of low-level inputs so that they are able to operate the counter. When a fairly high degree of amplification is provided some means of attenuating strong inputs is usual, or harmonic and other features of the input may add to the count. These cease to be recorded as the input is attenuated.

Counter Section

Figure 40 is a block diagram of this. It will require six LED numerals with holders, six 7447 decoder-drivers with holders, and six 7490 decimal dividers with holders.

Connections from the 7490 to the 7447, and from 7447 to the numeral, have been shown in earlier circuits. Each 7490 receives input at 14. Each 7490 (except for the last numeral) provides output at 11 for the next divider.

Each 7490 has 6, 7 and 10 grounded to negative. At each 7490 2 and 3 are connected together, and to 2 and 3 of the next 7490. This is the reset line, common to all six dividers. The

FIG. 40.

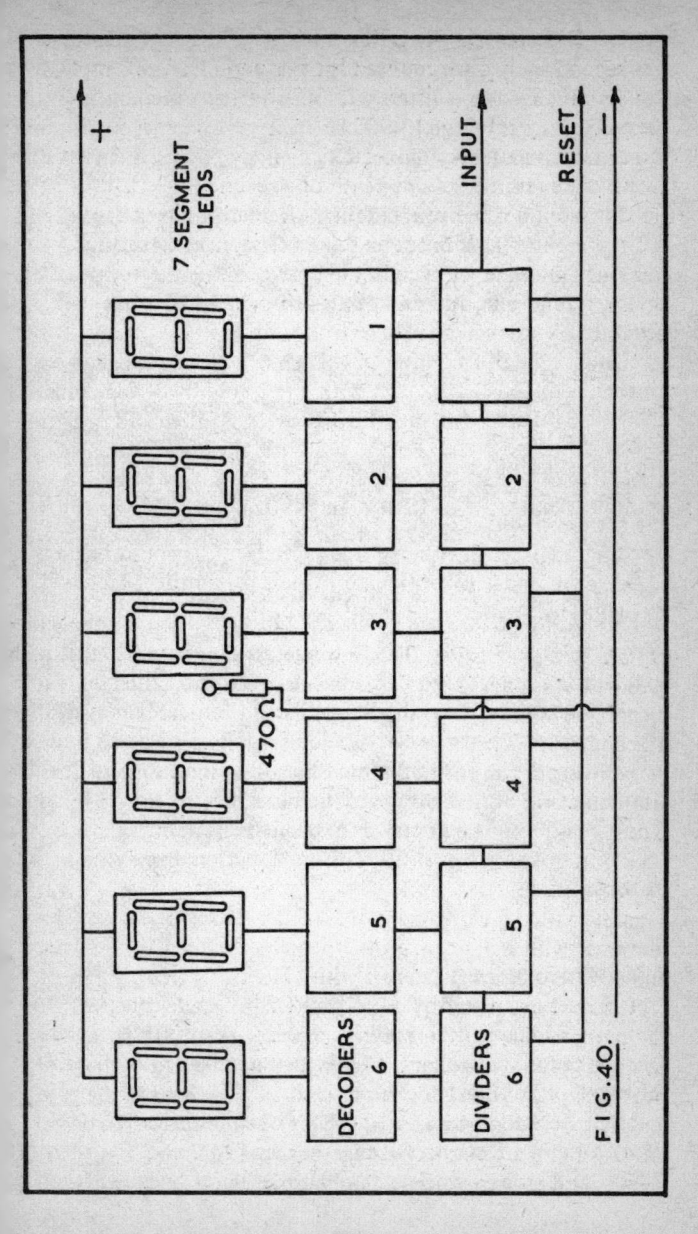

counter operates when this line is switched to negative. With this switch open, all numerals indicate zero.

The decimal point of the third numeral is permanently lit (pin 6) by current from a 470 ohm resistor. Positive and other supplies for the numerals and ICs are arranged in the way previously explained.

This section is constructed on 0.1in matrix board, and needs only four external connecting points. These can be colour-coded flying leads, or pins on the board. They are positive, negative, reset line, and input to 14 of the first or units divider.

Hand wiring of a counter of this kind is greatly speeded by using thin tinned copper wire with a few colours of very small diameter insulated sleeving. First provide positive and negative bus-bars of 20swg wire, for the ICs. It is then best to reproduce the wiring of one section for all sections, point by point. As example, connect 1 to 12 at each 7490 and run the wire on to 7 of its 7447. Connect a wire to 11 of each 7490, pass through sleeving to 6 of each 7447, solder and cut off. In this way the correct duplication of circuits is eased.

The finished board can be checked by temporarily connecting a 5V supply, and noting that all numerals are zero until the reset line is connected to negative. An input may then be taken to 14. This can be from the 555 pulser, at low frequency at first to observe correct working of units and tens, then speeded up to allow checking of hundreds and thousands until all numbers reach 9. A fault at any numeral may be traced to an IC, connections, or numeral, as explained elsewhere, and ICs are easily tested by substitution from a section displaying correctly.

Clock
Figure 41 is a block diagram of this. Here, a 1MHz crystal controlled oscillator is fitted. Subsequent decade dividers reduce this frequency to 100kHz, 10kHz, 1kHz, 100Hz and 1Hz, or 1 pulse per second. Pulses may be taken at the rate of either ten per second, or one per second. These go to the gate enabling section, opening the gate for one-tenth second, or one second, for two frequency ranges, as mentioned.

This section uses a crystal oscillator, as already shown, and

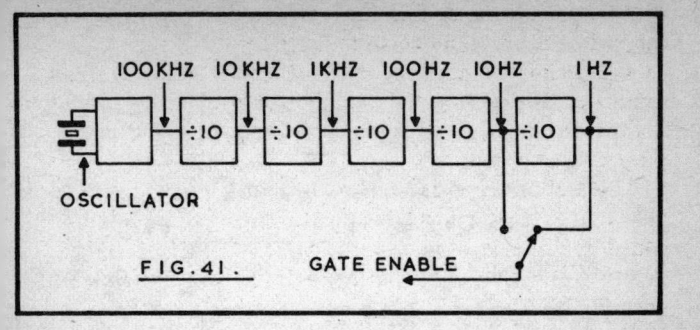

FIG. 41.

five 7490 decade dividers, with holders. Ground to negative 2, 3, 6, 7 and 10 of each 7490. Take input to 14, and output from 11. Also join 1 and 12 at each IC. A positive supply is required at 5 of each. By-pass positive to negative with a 0.1 µF capacitor on the board. Additional information on such a crystal oscillator, with wiring, will be found earlier.

The clock can be checked by means of the counter board, by taking 10Hz or 1Hz output to the board input (14). Join positive and negative lines. The count should advance at 10 or 1 per second, according to input frequency selected.

The crystal can be set exactly to frequency in the way explained for the radio frequency marker.

The possibility of a plus or minus error of 1 arises with frequency counters because there is no synchronisation between the incoming pulses to be counted, and clock pulses. This is normally ignored.

Gate & Reset

Two ICs are used for this section, Figure 42. The 7400 provides a noiseless initiating switch and gate to pass incoming pulses. The 7473 is a J–K flip-flop arranged so that the gate allows pulses to pass to the counter for one timing interval only, upon each occasion a reading is taken.

The operation of two NAND gates in the noiseless switch circuit has been described. It controls 5 of the 7473. Clock pulses at 1 second intervals (or some other chosen frequency) arrive at 1. When the 7473 is set by that half of the 7400 forming the noiseless switch, from 11, the next pulse at 1 of

FIG. 42.

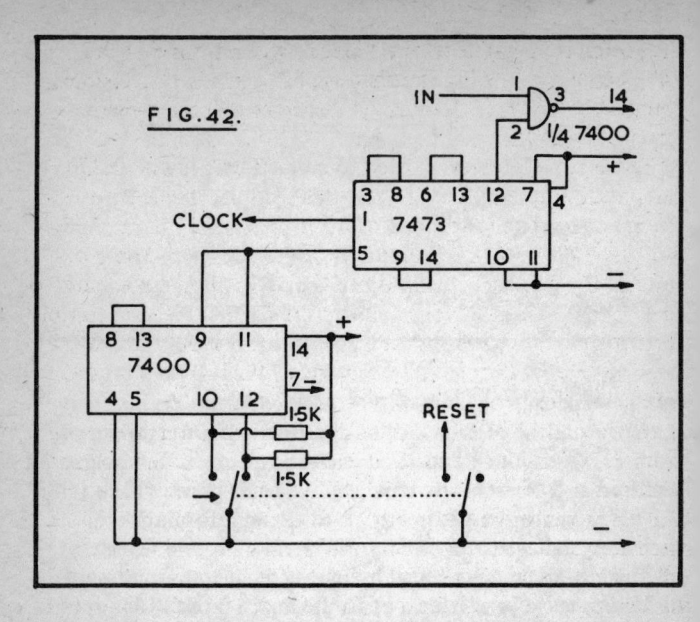

the 7473 gives output at 12 which allows the gate to conduct
when pulses arrive at the gate input 1 of the 7400. These
pulses, of the frequency being measured, thus appear at 3
of the gate. When the next clock pulse arrives at 1 of the 7473,
output from 12 closes the gate. Pulses at 1 of the 7400 gate
no longer appear at 3, and the count stops.

A frequency reading is obtained by pressing and releasing
the push-switch. The gate then opens at the next clock pulse,
the counter operates, and the gate closes, leaving the count
displayed.

As numerals need resetting to zero between counts, a second
pole on the push-switch opens the reset line, so that this is
obtained automatically. A 2-pole spring loaded leaf switch is
most suitable, and contacts are bent so that the reset line is
completed before counting starts. It should be noted that the
connections from the noiseless switch are not reversible, as if
the gate is enabled while the reset line is open, no indication can
arise. Reversed connections here can also cause different
counts each time, due to the reset line only being closed for a
part of the counting interval, which will in turn depend on how

the switch is operated. A mechanical switch from 5 of the
7473 can be reasonably satisfactory, but can result in changing
counts with the same frequency, due to switch imperfections, as
explained elsewhere.

These two ICs may be fitted to a board which will also
carry the input amplifier. To test this section, connect 1
second pulses from the clock board to 1 of the 7473, and take
3 of the 7400 gate to the input 14 of the counter. Apply a
suitable input — such as that from the 555 pulser shown later —
to 1 of the 7400 gate. The counter should then be seen to run
for 1 second, after the switch is pressed, and should display the
same number each time (plus or minus 1) unless the input
frequency changes (due to temperature changes, etc.).

The switch should be located for easy operation, when the
counter is in use. This method leaves a stable reading, which
will remain as long as wanted. The switch is pressed each time
a new reading is required. For automatic operation it is
necessary to have a multivibrator (Figure 44) or other means
of controlling the 7473, giving suitable repetitive count and
display times. To provide a continuous display, the count has to
be taken into binary coded memories or latches, then to the
decoder-drivers, with a control section which initiates each
action in correct sequence.

For two ranges, a 2-pole 2-way switch can be used to take
clock pulses at 1 second and 0.1 second intervals.

Amplifier

A simple amplifier which will provide a sensitivity of approximately 0.5 volt is shown in Figure 43. VR1 allows attenuation
of greater inputs where necessary. The transistor collector swings
negative when the transistor is conducting. Pulses only pass
through the gate to be counted when this is opened in the way
described.

A screened lead with earthing in the usual way is generally
necessary to avoid mains frequencies becoming associated
with the input. For high freuqencies, stray capacitances must
be kept down, both in layout and test lead, or sensitivity falls
off from the shunting effect introduced in this way.

An increase in frequency, beyond that with which a counter
is normally able to deal, is usually provided by means of a pre-

FIG. 43.

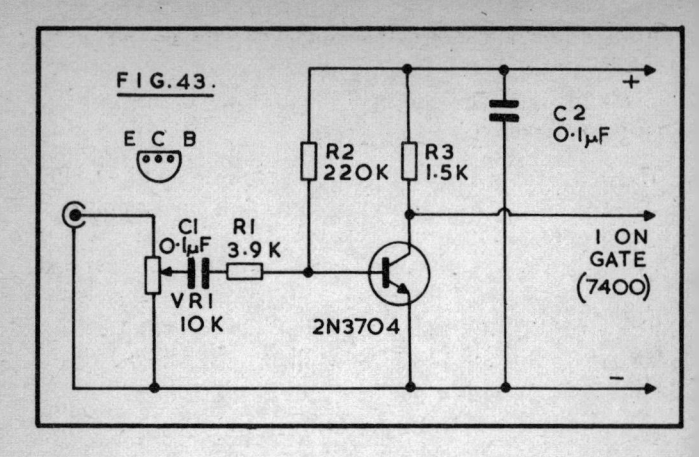

E C B

C1 0.1µF
VR1 10K
R1 3.9K
R2 220K
R3 1.5K
C2 0.1µF
2N3704
1 ON GATE (7400)

scaler, if required. Typically, this can consist of a single decade divider, able to operate at high frequency. It is fed from the frequency to be measured, and its output passes to the counter. As it divides by 10 (or other chosen figure) all the frequency counter indications are multiplied by 10.

Increased sensitivity, where wanted, is obtained by using a further amplifier, and a Schmitt trigger is frequently placed in the amplifier output circuit, before the gate. This is operated by the sine and other waveforms present. A variable attenuator (VR1) may be replaced by a switched resistor attenuator, with a number of input levels.

Where required, DC isolation of input is obtained by placing a capacitor in series with the input circuit. This can limit the low frequency, or prevent very low speed counts. Diode limiters may be present across the input, to clip this to a preset level. Various methods of providing compensation for changes in frequency can also be provided.

Errors in counter readings are most likely to arise from external sources, such as pick up or leak-through of mains frequencies or strong RF or other frequencies. Overloading may allow otherwise insignificant harmonics to register. Insufficient level of input may result in reduced counts if the level varies, only some pulses being of sufficient magnitude to operate the counter. Very high numeral indications, with no input, usually indicate instability in the amplifier section, which is generating oscillations which pass the gate and operate the counter section.

DIGITAL SIGNAL GENERATOR

This uses a counter with five numerals to read up to 99.999, clock, gate, and gate operating circuits with input amplifier as shown for the frequency counter. To enable the gate to pass the signal from the signal generator section in order that the frequency can be displayed, 5 of the 7473 must go low, then the count continues for one clock pulse when 5 goes high. This is initiated by one-half of the 7400, using two gates, as shown.

Automatic display can be provided by means of the multivibrator in Figure 44. C1 and C2 are of dissimilar value. The collectors of Tr1 and Tr2 drive the 7400. Point 10 goes high briefly, and so does the counter reset line, for return to zero. At this time 12 is low. Point 10 and the reset line are held low for a longer period. This allows the 7473 to open the gate at the next clock pulse, and close it at the following clock pulse, the count remaining displayed.

In use, frequencies are displayed for about 2 to 3 seconds, this being repeated automatically. The count-up for 1 second periods is readily seen, but for 0.1 second or shorter intervals can scarcely be seen, and neither interferes with the easy

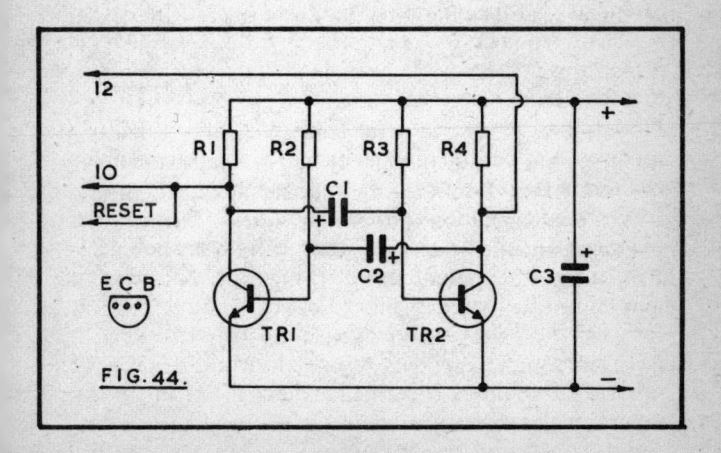

FIG. 44.

reading of the number shown.

Note that 12, and 10 (with reset line) are not reversible, unless C1 and C2 should also be exchanged. R1 and R4 may be 1.5k each in parallel with the similar resistors in Figure 42, if this auto-display is added.

Components for Figure 44
 (Resistors 5% ¼W)

R1, R4	750 ohm	C2	2 μF
R2, R3	33k	C3	470 μF
C1	470 μF	Tr1, Tr2	2N3706

With the multivibrator controlling the counter, results should be the same as when the reading was obtained by means of the 2-way push-switch, except that they are obtained, displayed and cancelled automatically.

The upper frequency limit at which such a counter will operate depends largely on the first decade divider. Where a high frequency is required, this is generally a selected IC, an IC designed for HF purposes, or a pre-scaler as described.

Where the waveform is not important, the 555 pulser can be arranged to operate up to about 50kHz by employing a switch with more ways, selecting capacitors down to 2,000pF. The output may operate the counter for frequency display directly, or may feed the amplifier through a resistor of about 8.2k.

Audio Oscillator

Figure 45 is the circuit of an adjustable audio oscillator able to operate over the frequency range of approximately 15Hz to 20kHz. Output is amplitude stabilised, and of sine wave quality. Operation is from a 9V supply. The counter input amplifier is fed via a series resistor from the point C, the input amplifier potentiometer in Figure 43 not being required.

FIG.45.

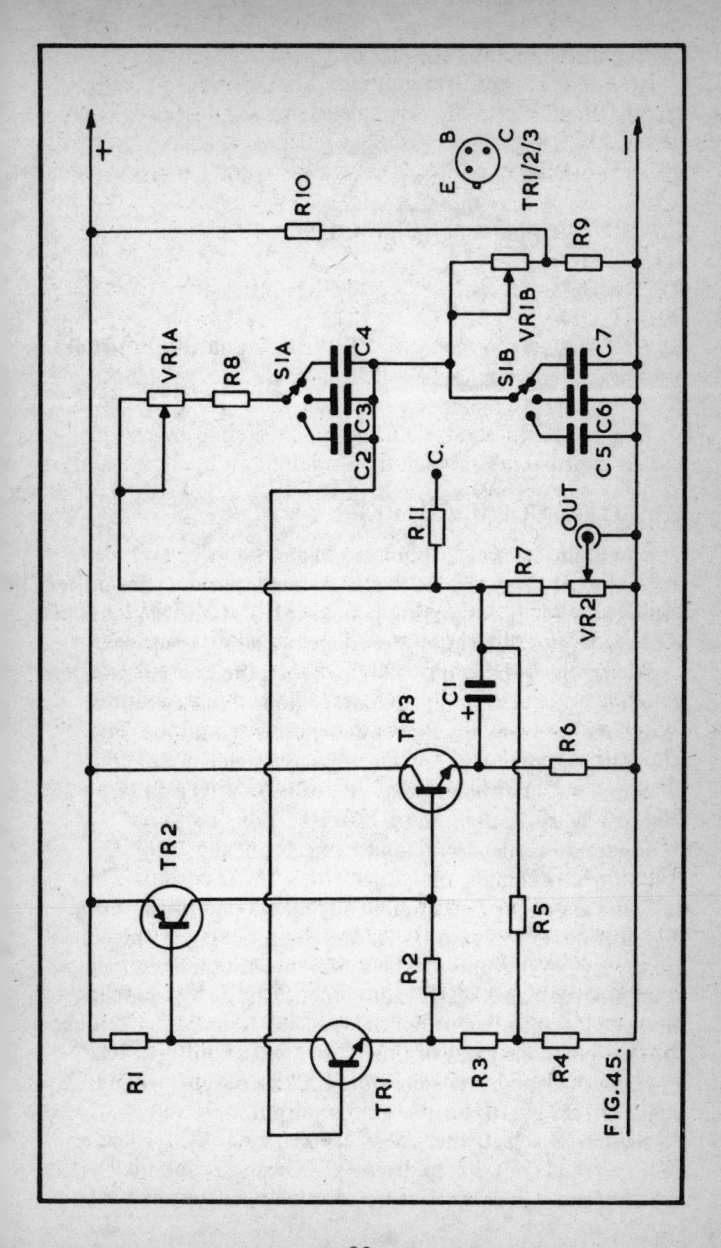

Components for Audio Oscillator (Figure 45)
(Resistors 5% ¼W)

R1, R2	6.8k	VR2	1k linear potentiometer
R3	1.2k		
R4	100 ohm	C1	1000 µF 10V
R5	Type R53 thermistor	C2, C5	1 µF
R6	680 ohm	C3, C6	0.1 µF
R7	10 ohm	C4, C7	10nF
R8, R9	1.8k	Tr1, Tr3	BC108
R10	10k	Tr2	BC186
R11	8.2k	S1A/B	2-pole 3-way switch.
VR1A/B 2-gang 10k linear potentiometer		Board, etc.	

RADIO FREQUENCY MARKER

This unit allows exact calibration of a radio receiver, signal generator, or the variable frequency oscillator of a transmitter. Provision is made for marker pips at 1MHz, 100kHz and 10kHz intervals, up to 30MHz; or higher with a sensitive receiver.

In use, the 1MHz pips will only be required with equipment which is badly out of adjustment, or home-built apparatus where the frequency is almost completely unknown. The 100kHz pips appear at one-tenth intervals, and are of considerable use. For bandspread receivers or VFO calibration, the 10kHz pips allow division of 100kHz sectors into ten.

Figure 46 is the circuit, and two gates of the 7400 IC1 function as oscillator, operating with a 1MHz crystal. The resistors are to provide suitable operating conditions. Trimmer T1, in conjunction with C1, allows slight pulling of the crystal frequency, so that this can be set by means of an external standard, such as 200kHz radio broadcasts, or the standard frequency transmissions on 2.5MHz, 5MHz and other frequencies.

The remaining gates of this IC are used as buffers. IC2 provides division by 10, and thus 100kHz output, while IC2 again divides by 10, for the 10kHz output.

Switch S1 selects the 1MHz, 100kHz or 10kHz signal, as required, and uses this to drive Tr1. Output is obtained at the emitter, and signal level can be adjusted by means of VR1.

FIG.46.

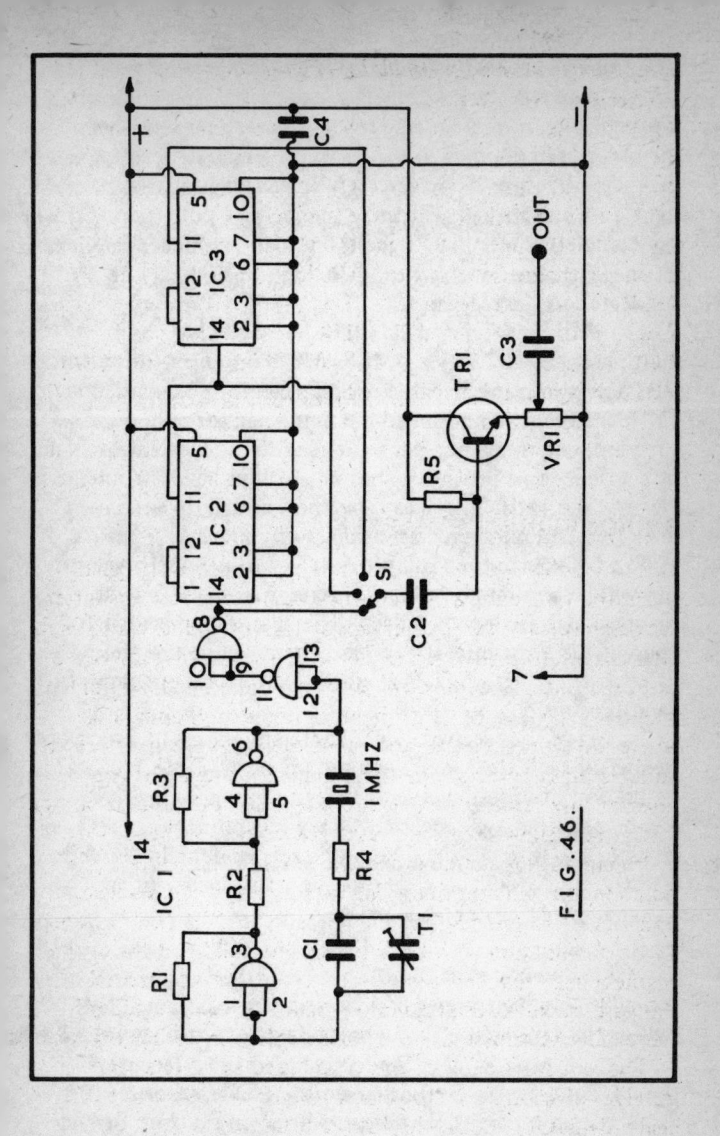

Output Use

For many purposes, sufficient input to the receiver will be obtained if a short lead from the marker output socket is run near the aerial socket of the receiver, or near a short wire taken to the receiver socket. The usual aerial is disconnected. Coupling may need adjusting to suit the receiver sensitivity and the order of harmonics, which grow progressively weaker, until they can no longer be detected. VR1 can be turned back where signals are too strong. The usual inexpensive potentiometer will not completely attenuate signals, especially at higher frequencies. When signals are not too strong, the output lead may be connected to the receiver.

The 1MHz marker signals will be heard at 1MHz and multiples — 2MHz, 3MHz, 4MHz, and so on, throughout the receiver tuning ranges. Scales or dials can thus be marked at 1MHz intervals. With home-built equipment of unknown coverage, one frequency has to be identified on each waveband, by reference to a broadcast station, amateur band, or similar means. The 1MHz marks can then be counted up and down from this, and numbered according to frequency.

The 100kHz output will provide 10 pips per 1MHz sector, and will be appropriate for calibration over many general coverage high frequency ranges. The 10kHz marker signals again divide these into 10 sectors. These will be too close together for many general coverage receivers, except those with bandspread tuning.

Tuning of the receiver may be noted by means of a beat frequency oscillator, which will usually be available as exact calibration of this kind is most likely to be wanted with a communications type receiver. A tuning indicator can also be used, and may consist of a multi-range meter clipped on to read the anode, collector, emitter, or drain current of an automatic volume controlled stage.

Crystal Trimming

An extremely high degree of accuracy can be obtained by pulling the crystal into frequency with an external source. To do this, use an aerial with the receiver, to obtain reception of the standard frequency transmissions on 2.5MHz, or receive 200kHz transmissions. Differences between the 25th or 2nd harmonic of 100kHz and this signal will be heard as a growl or flutter, which will fall in frequency as T1 is adjusted in the correct direction. For best results and easy adjustment, signals from the station tuned in, and the marker generator, need to be of roughly similar strength, or at least not so different that one swamps the other.

VFO Calibration

To calibrate a VFO, couple the marker output and the VFO to a receiver. Tune in a marker harmonic with the receiver, and tune the VFO to zero beat with this, and calibrate the VFO dial to suit. Continue as necessary. As example, at 3.5, 3.6, 3.7 and 3.8MHz for the 80m band, subsequently filling in with 10kHz points.

Accuracy of receiver tuning does not influence calibration. The receiver is merely used to compare the frequencies of the VFO and wanted harmonic.

To calibrate a signal generator, proceed in the same way, except that marks will generally be at wider points, except on low frequencies.

Correct calibration will allow the easy location of high frequency bands, stations using known frequencies or wavelengths, etc. Trimming and alignment of calibrated receivers may be directed towards securing the best agreement between dial readings and actual frequencies throughout.

Components for Radio Frequency Marker (Figure 46)
   (Resistors 5% ¼W)

R1	1.8k	IC1	7400
R2	220 ohm	IC2	7490
R3	560 ohm	IC3	7490
R4	220 ohm	3 off 14 pin DIL holders	
R5	470k	T1	30pF trimmer
VR1	1k linear pot	Tr1	2N3706
C1	22pF	S1	Single pole 3-way
C2	220pF	1MHz	crystal & holder
C3	220pF	Board, etc.	
C4	0.01 μF		

VARIABLE 555 PULSER

When making or testing a counter, whether decimal, or for seconds, minutes and hours, some form of rapid checking is useful. A variable pulser, using the 555 integrated circuit, will be convenient for this. Four ranges are fitted, for approximately 1 pulse per 20 seconds to 3Hz, 3Hz to 160Hz, 160Hz to 5,000Hz, and 5kHz to 200kHz.

The slow pulses allow observation of the seconds or first digit display of a counter or timer, and individual testing of various digital circuit points for high and low, if required. To speed up checking in tens, hundreds and other sections (or minute or hour sections) pulses can be speeded to run the whole counter at a greatly increased rate. In this way, as example the whole numeral sequences of a 24 hour clock can be observed in a few minutes.

Figure 47 shows the circuit. It may be operated from the same 5V supply operating the counter or timer, or from a

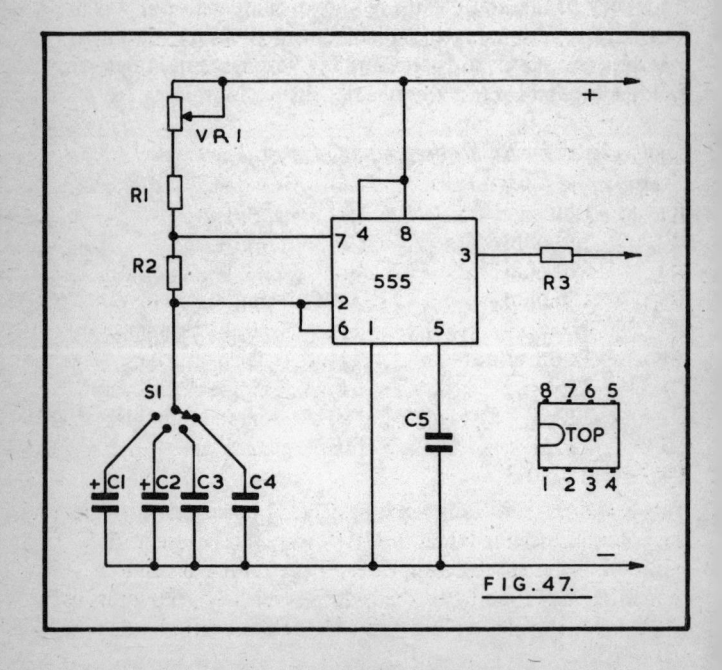

FIG. 47.

4.5V (3-cell) battery. If a separate supply is used, join negatives of pulser and counter or timer.

VR1 provides adjustment of frequency. S1 selects one of a range of timing capacitors. The larger values provide the lower rates. It is, of course, not essential that the ranges be exactly as mentioned, so other capacitor values, or modification to VR1, would be possible.

Output is from 3 of the 555, via a resistor which limits fault current in the event of a defect or wrong connection.

The few components are readily wired on a small 0.1in matrix board. This, and S1 and VR1, can fit inside a case or box. Red and black flexible leads, with small clips, will allow easy connecting up to the power lines of the timer or counter which is to be checked. The output lead can be fitted with one of the miniature insulated prods with clip end operated by a button, and this can then be attached to any required point, or to 14 on the binary coded decimal dividers or other ICs.

A check of the pulser should show a range of rates, adjustable in frequency to suit the purpose, and according to the position of S1. Current drain is small, typically 2mA to 3mA or so.

Components for Variable 555 Pulser (Figure 47)
  (Resistors 5% ¼W)

R1	4.7k	C3	25nF
R2	2.2k	C4	300pF
R3	180 ohm	S1	single-pole 4-way switch
VR1	500k linear potentiometer		
C1	33 μF	555 and 8-pin DIL holder	
C2	1 μF	Knobs, case, board, etc.	

1-ARMED BANDIT

This is a more ambitious game project, but can provide a great deal of amusement. When the RUN switch is pressed, three numbers change at different rates, those arising at random remaining displayed when the switch is released. To improve chances for the player, individual HOLDS are fitted, so that

FIG. 48.

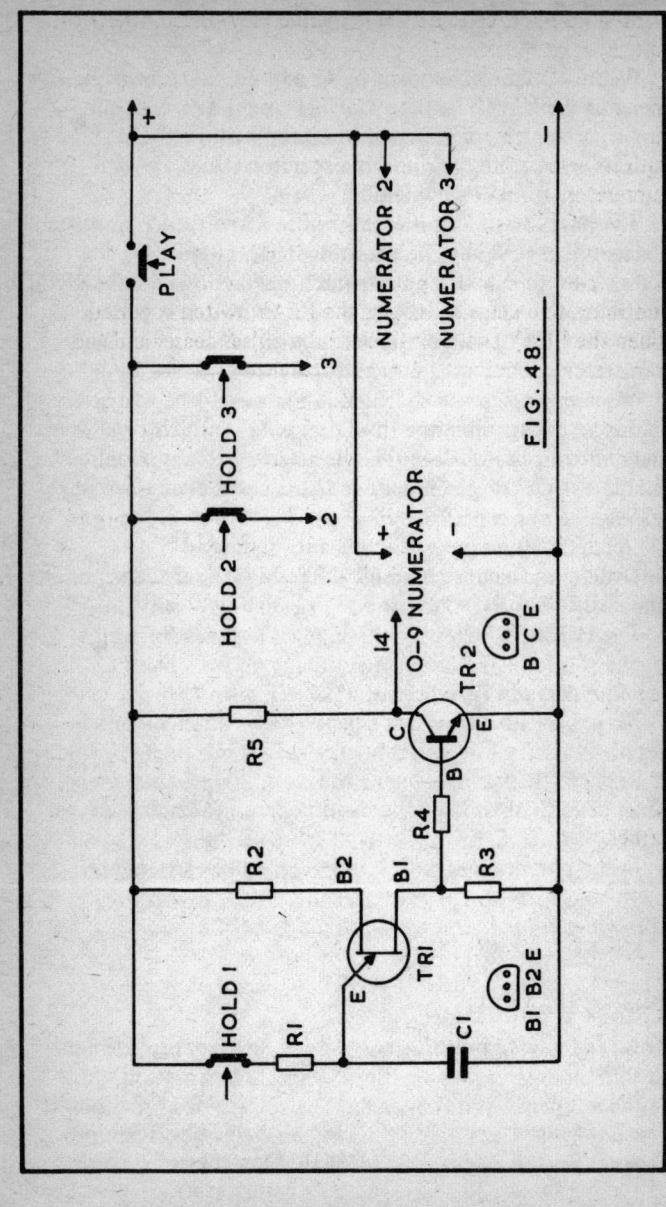

wanted numbers can be retained while playing for others to suit.

Figure 48 shows the circuit. R1 and C1 control the pulsing speed of the UJT Tr1. Base 1 of Tr1 drives Tr2, which in turn operates the numerator from its collector circuit. Tr1, Tr2, with its associated items and the switch HOLD 1, with 0—9 numerator, form one numeral.

The HOLD switch is normally closed, and PLAY switch is normally open. When the PLAY switch is closed, Tr1 commences to produce pulses which operate this numeral, which continues to change so long as the PLAY switch is closed. When the PLAY switch is opened the pulses cease, and the numerator continues to display the figure.

The second figure is obtained in the same way, and has HOLD 2 switch, while the third figure also operates in this manner, with HOLD 3 switch. Should any of the switches HOLD 1, HOLD 2 or HOLD 3 be kept open, then no voltage reaches the pulser section, when the PLAY switch is closed, so the related numeral remains unchanged.

Different values are used for C1 and R1, so that the numbers run at different rates.

The HOLD switches should be push-to-break, or spring-loaded toggle types which are normally closed. The PLAY switch is also of this type, but normally open.

All pulser and numerator negatives are common to the negative line. HOLD switches may switch R1, or R1, R2 and R5 together, without changing the count. Numerals can be Nixie or LED, with the dividers, decoders and circuits shown earlier.

In the component list, R1 and C1 are given separately.

Components for 1-Armed Bandit, (Figure 48)
(Resistors 5% ¼W)   3 off:

R1a 47k   C1a 0.1 µF   R2   220 ohm
R1b 47k   C1b 0.22 µF   R3   680 ohm
R1c 220k   C1c 0.47 µF   R4   1k
                         R5   1k

PLAY switch   Tr1   UT46, TIS43, E5567
Board   Tr2   2N3704
Case   HOLD switches
                         0—9 Numerators.

Code	Title	Price
160	Coil Design and Construction Manual	1.25p
202	Handbook of Integrated Circuits (IC's) Equivalents & Substitutes	1.45p
205	First Book of Hi-Fi Loudspeaker Enclosures	95p
207	Practical Electronic Science Projects	75p
208	Practical Stereo and Quadrophony Handbook	75p
211	First Book of Diode Characteristics Equivalents and Substitutes	1.25p
213	Electronic Circuits for Model Railways	1.00p
214	Audio Enthusiasts Handbook	85p
218	Build Your Own Electronic Experimenters Laboratory	85p
219	Solid State Novelty Projects	85p
220	Build Your Own Solid State Hi-Fi and Audio Accessories	85p
221	28 Tested Transistor Projects	1.25p
222	Solid State Short Wave Receivers for Beginners	1.25p
223	50 Projects Using IC CA3130	1.25p
224	50 CMOS IC Projects	1.25p
225	A Practical Introduction to Digital IC's	1.25p
226	How to Build Advanced Short Wave Receivers	1.20p
227	Beginners Guide to Building Electronic Projects	1.25p
228	Essential Theory for the Electronics Hobbyist	1.25p
RCC	Resistor Colour Code Disc	20p
BP1	First Book of Transistor Equivalents and Substitutes	60p
BP2	Handbook of Radio, TV & Ind. & Transmitting Tube & Valve Equiv.	60p
BP6	Engineers and Machinists Reference Tables	70p
BP7	Radio and Electronic Colour Codes and Data Chart	35p
BP14	Second Book of Transistor Equivalents and Substitutes	1.10p
BP23	First Book of Practical Electronic Projects	75p
BP24	52 Projects Using IC741	95p
BP27	Giant Chart of Radio Electronic Semiconductor and Logic Symbols	60p
BP28	Resistor Selection Handbook (International Edition)	60p
BP29	Major Solid State Audio Hi-Fi Construction Projects	85p
BP32	How to Build Your Own Metal and Treasure Locators	1.35p
BP33	Electronic Calculator Users Handbook	1.25p
BP34	Practical Repair and Renovation of Colour TVs	1.25p
BP35	Handbook of IC Audio Preamplifier & Power Amplifier Construction	1.25p
BP36	50 Circuits Using Germanium, Silicon and Zener Diodes	75p
BP37	50 Projects Using Relays, SCR's and TRIACs	1.25p
BP38	Fun and Games with your Electronic Calculator	75p
BP39	50 (FET) Field Effect Transistor Projects	1.50p
BP40	Digital IC Equivalents and Pin Connections	2.50p
BP41	Linear IC Equivalents and Pin Connections	2.75p
BP42	50 Simple L.E.D. Circuits	95p
BP43	How to Make Walkie-Talkies	1.50p
BP44	IC555 Projects	1.75p
BP45	Projects in Opto-Electronics	1.25p
BP46	Radio Circuits Using IC's	1.35p
BP47	Mobile Discotheque Handbook	1.35p
BP48	Electronic Projects for Beginners	1.35p
BP49	Popular Electronic Projects	1.45p
BP50	IC LM3900 Projects	1.35p
BP51	Electronic Music and Creative Tape Recording	1.25p
BP52	Long Distance Television Reception (TV–DX) for the Enthusiast	1.95p
BP53	Practical Electronic Calculations and Formulae	2.25p
BP54	Your Electronic Calculator and Your Money	1.35p
BP55	Radio Stations Guide	1.75p
BP56	Electronic Security Devices	1.45p
BP57	How to Build Your Own Solid State Oscilloscope	1.50p
BP58	50 Circuits Using 7400 Series IC's	1.35p
BP59	Second Book of CMOS IC Projects	1.50p
BP60	Practical Construction of Pre-amps, Tone Controls, Filters & Attn	1.45p
BP61	Beginners Guide to Digital Techniques	95p
BP62	Elements of Electronics – Book 1	2.25p
BP63	Elements of Electronics – Book 2	2.25p
BP64	Elements of Electronics – Book 3	2.25p
BP65	Single IC Projects	1.50p
BP66	Beginners Guide to Microprocessors and Computing	1.75p
BP67	Counter Driver and Numeral Display Projects	1.75p
BP68	Choosing and Using Your Hi-Fi	1.65p
BP69	Electronic Games	1.75p
BP70	Transistor Radio Fault-Finding Chart	50p
BP71	Electronic Household Projects	1.75p
BP72	A Microprocessor Primer	1.75p
BP73	Remote Control Projects	1.95p
BP74	Electronic Music Projects	1.75p
BP75	Electronic Test Equipment Construction	1.75p
BP76	Power Supply Projects	1.75p
BP77	Elements of Electronics – Book 4	2.95p
BP78	Practical Computer Experiments	1.75p
BP79	Radio Control for Beginners	1.75p
BP80	Popular Electronic Circuits – Book 1	1.95p
BP81	Electronic Synthesiser Projects	1.75p
BP82	Electronic Projects Using Solar-Cells	1.95p
BP83	VMOS Projects	1.95p
BP84	Digital IC Projects	1.95p
BP85	International Transistor Equivalents Guide	2.95p
BP86	An Introduction to Basic Programming Techniques	1.95p
BP87	Simple L.E.D. Circuits – Book 2	1.50p
BP88	How to Use Op-Amps	2.25p
BP89	Elements of Electronics – Book 5	2.95p